Marion Harland

The Home of the Bible

What I Heard and Saw in Palestine

Marion Harland

The Home of the Bible
What I Heard and Saw in Palestine

ISBN/EAN: 9783337278984

Printed in Europe, USA, Canada, Australia, Japan

Cover: Foto ©berggeist007 / pixelio.de

More available books at **www.hansebooks.com**

AN ORIENTAL BARBER-SHOP—A FAMILIAR STREET SCENE IN DAMASCUS.

Home of the Bible

What I Saw and Heard in Palestine

An account of the Sacred Places, Battle scenes, miracle-haunted hamlets and holy Homes of the country of David and the Christ, together with sketches of Historical events, tragedies and romances, marvelous legends, customs and characteristics, hopes and promise of the Race of Israel.

"Put off thy shoes from off thy feet; for the place whereon thou standest is holy ground."

By Marion Harland

Author of

"Common Sense in the Household," "The Royal Road," "A Gallant Fight," "His Great Self," etc.

Copiously Illustrated with engravings from photographic views taken in Palestine, to which has been added

The Story of Martyred Armenia

The Christian People of Ancient Eden, and their Cruel Persecutions by the Moslems

By G. H. SANDISON

Associate Editor of "The Christian Herald"

THE CHRISTIAN HERALD

LOUIS KLOPSCH, Proprietor

BIBLE HOUSE, NEW YORK CITY

1896

THE VESTIBULE.

THIS book is not a diary of travel, drawn out in detail to fill a given number of pages. Nor does it deal with statistical data and archæological research. My object in going to Palestine was to see with fresh eyes, and judge with unprejudiced mind what this, the most interesting of lands to the Christian, is like to-day; how the inhabitants look, and live, and think, and what traces the traveler finds in the ruins of the Past of the nations who flourished, battled and were, in turn, swept from home and country by the march of GOD'S providence.

As a pilgrim, an observer of the people, a student of the various scenes, and a historian of the heart-stirring incidents which have immortalized for all the ages the country divided between, and apportioned to, the sons of Jacob, I visited the Home of the Bible. In reverence of spirit I pursued the path of John the Baptist in the wilderness, stood upon the spot where he baptized the "Greater than he," and looked across the Dead Sea to the ruins of Machærus where he was beheaded. In more profound reverence I trod the route taken by the Master during the three years of the ministry that began at "the place where John was baptizing" at the Jordan, and culminated upon Calvary. With a strange nearness of heart and thought to the people and times of patriarch, prophet and apostle, I have looked upon the homes of Abraham, David, Isaiah, Samson—of the disciples of Christ and the Mother who bore Him.

Of these places and so many more that the enumeration here would be tedious, I have written in familiar style, avoiding statistics and dry historical details—most of all, moralizing and preaching. In short, I have let place and people speak for themselves.

The people of Palestine have changed less in their manner of living, customs, and their prejudices, than those of any other country; they are almost the same to-day that they were two thousand and more years ago. I therefore saw them practically as Jesus observed them, and studied their characters by the light of history. I watched workmen in their shops, husbandmen in the fields, congregations in the synagogues, and women in their homes, that I might become familiar with the Jews, the Samaritans, the Gadarenes, and all that remnant of ancient Israel whose lives span the chasm of years between Abraham and the present. I have written for the masses, and if my readers enjoy these contributions to the modern history of Palestine as much as I enjoyed the preparation of the same, I shall esteem my labors well rewarded and my purpose completely fulfilled.

The journey was a delight. The telling of it has been a joy into which I would fain admit the many who, not having seen the Holy Land, yet love it, and to whom whatever will set her clearly and truthfully before their mental vision will come as good news from a far country.

Marion Harland

THE STORY OF ARMENIA.

ILLUSTRATIONS

The Home of the Bible.

CHAPTER I.

FROM PORT SAÏD TO JAFFA.

WE had expected to rough it at some, and probably at several, periods of our journeyings. We had not expected that the process would begin upon one of the far-famed steamers of the Peninsular and Oriental line, plying between England, Australia and India. Perhaps we were unreasonable in comparing "The Coromandel," on which we took passage from Marseilles, with the floating palaces that make an Atlantic voyage endurable to the least sea-worthy tourist. Certain it is that the Peninsular and Oriental craft suffered grievously by the contrast. The table was little better than that of a second-class boarding-house; the appointments of the state-rooms—"cabins" as they are called by the English—were mean and uncomfortable, the dining-saloon so narrow and dark as to make meals a penitential process.

Added to these drawbacks to the comfort of the voyagers was the circumstance that the Mediterranean was in a bad humor, and so maltreated us that three-quarters of the land-farers on board were wretchedly ill, until we were flung, as it were, into the Straits of Messina, and, under the lee of Sicily, glided into the blue placidity of what, for the first time since our embarkation, deserved the name of a "summer sea." It kept up this reputation to the end of our voyage—the harbor of Port Saïd, at the mouth of the Suez Canal. There we discovered, to our chagrin, that the steamer which was to have taken us on up to Beirût upon the day succeeding our arrival was detained—nobody knew where—in quarantine—nobody knew for what—and no other would sail within four days. The town is drearily modern, cheaply built as an entry-port to the canal upon a low-lying sand-bank.

Before we had been an hour in the "Hotel Continental," we made the discovery that there was not a woman-employé in the house. Our beds were made, our rooms swept and dusted, and all other functions of chambermaids performed by men, brown of face, black-eyed, and arrayed in a mongrel costume of jacket, blue blouse and trousers. They moved quickly and quietly, they answered the bell promptly and were the very soul of civility. Recalling certain experiences in what we used to call "the waiter-girl belt" of the Western States, and the deliberate indifference, sometimes verging upon sullenness, of the "young ladies out at service" in the Middle and Eastern sections of our free and enlightened land, we confessed that the chamberman system has notable advantages.

THE ROADSTEAD AT JAFFA.

During our first stroll through the little town, we wondered where all the women who must wive the men, and mother the boys thronging the streets kept themselves. Except for an occasional French, or English, or American, in bonnet, jacket and gown of European birth or descent, we met not one representative of "the dominant sex." On the next day, we took one of the few carriages that are a discord in the Oriental tone of the region, and, accompanied by a swarthy dragoman from the hotel, drove through the native quarter.

PORT SAÏD MATRON. (35)

Women were there, in number and variety sufficient to quiet our speculations. They sat flat in the doors of the houses, or upon the ground against the outer walls of the same, knitting, tending babies and gossiping; they stirred messes in brass pots over braziers of charcoal, and made coffee for men who sat cross-legged upon the bare earth to drink it; they sold beads, sweetmeats, sausages, uninviting vegetables and less attractive fruit—apples, dates and lemons—at stalls set right in the street. However engaged, they were covered up to the eyes in a sort of combination-suit of circular cloak, hood and veil, usually black, and of some opaque material. It is picturesque, and, after the manner of many other picturesque costumes, it must be inconvenient and unpleasant, especially in a country where the thermometer in December ranges at noon from seventy-three to ninety-six in the shade. How a native woman, thus muffled up and swathed, can discharge even the few domestic duties incumbent upon one who, as we have seen, lives and keeps house mainly in the open air, is a hopeless puzzle to the foreign observer. That nothing may be lacking from the discomfort of mantle, wimple and veil, our Arabian, or Egyptian, or Syrian matron wears, in token of her honorable estate of wedlock, a brass tube, nearly an inch in diameter, strapped perpendicularly upon her forehead directly between her eyebrows and in a line with the bridge of her nose. Close scrutiny showed us the black string attached to one end and losing itself in the close black hood which forms the upper part of the mantle; how it was attached to the veil below is still to us an Oriental enigma. Through this pipe, Mohammed, or some other Moslem authority in spiritual matters, is supposed to breathe admonition, counsel and consolation as the married devotee requires it.

Making our way slowly through the motley crowds of the native quarter, our coachman and dragoman yelling incessantly to the pedestrians to clear the roads, and the whip of the driver doing sharp execution upon the bare legs of innumerable boys and lads whose brown bodies were covered to the knee by a single garment, half-frock, half-trousers—we at length escaped from din and dinginess and a nameless and altogether nauseous mixture of vile odors, into an open roadway, laid along the water's edge to the cemetery, two miles from the town. To secure this roadway, a curbing of solid stone was laid on both sides, the space between being filled with earth and sand. After leaving the outskirts of Port Saïd, there were no signs of human habitation except a few scattered hovels dotting the waste on our left. About the doors small black pigs rooted and squealed; a stray dog skulked in the forlorn hope of a supper, and toward the desert three camels, with outstretched noses, followed their masters. We watched them kneel to be unloaded, and then remain quiet on the sand for their nightly rest. The only symptom of vegetation, as far as eye could reach, was in clumps of something we mistook for beach-grass, until the dragoman plucked a branch for our inspection, and we found it strangely succulent, with fleshy leaves and even small yellow berries, ovate and pulpy.

Right in the middle of this desert arose the walls of the cemeteries—the Christian, devoted to the interment of French Roman Catholics and Egyptian Copts, and the Moslem, where lie "the faithful" of whatever nationality. We looked into the first, seeing nothing very different from the tall crosses and head-stones hung with tawdry immortelles and beads, such as we had beheld in dozens of other foreign burial-grounds. We alighted at the gate of the Moslem cemetery and entered the enclosure. Arid sand for many feet downward is the substance through which the graves are sunk. Within a few days after the mound is heaped above the sleeper below, the meeting winds of sea and desert tear it down and whirl the sand to the four quarters of the enclosure. Hence, as soon as may be, a box, of the shape and size of the grave, is fitted over it. When the relatives can afford it, a structure of similar form in cement takes the place of the wooden case. Upon box and cement are written the names of the deceased and texts from the Koran. Above many of the rude tombs arise coop-like constructions, with trellised sides and tops, within which stand pots of dwarf palms, of cacti, geraniums, and, once in a great while, of sickly vines, pathetic to behold in a region where rain does not fall for months together, and water is sold to the poor by the jar or skinful.

There are no regular walks or avenues, and wherever the graves were not protected by boxes the sand bore the imprint of many feet. Leading the way to the outermost row of graves, the guide pointed to a line of freshly-heaped mounds, to the head-boards of which were tied shabby bunches of palm-leaves, palm-branches and artificial flowers.

"If you had been here this morning," he explained in execrable French enlivened by insupportable English, "you would have seen two thousand women —perhaps more, maybe less—here, crying, and crying, and crying, and telling how good her husband was, or her child was so sweet, or how she mourned her father, or her sister, or her brother, and did break her heart for her mother, died so long ago. They come so, every Friday, and cry just the same and ever so hard, and it is they who do this"—pointing to the newer mounds. "These are they who were buried of late, you comprehend; and the ladies, they keep them high until the boards go about them—so they be not blown away by the sea-wind —you comprehend?"

Friday is the Moslem Sabbath, and this pious pilgrimage is a duty to be performed upon the holy day. Hearing the tale, we looked with different eyes upon the sandy heaps raised by pitying hands; the already withering memorials lashed to the main head-boards had meaning and poetry. The woman-heart is the same the world around.

(38) PORT SAÏD FAMILY COMING HOME FROM MARKET.

At the end of the row of new mounds was an open grave. "When is this to be filled?" I asked, knowing that burial in this tropical country follows with awful rapidity upon death, and supposing that the pit was dug purposely for somebody.

The dragoman shrugged his shoulders.

"Ah! who can know? It may be to-morrow; it may be next week. But there is always one ready. Somebody must come to fill it some day. You comprehend?"

Comprehend! ah, but too well! That, also, was an old, old story, known wherever men and women live and die.

Heaven forbid that we, or any of our blood, should ever die at Port Saïd!

I was awakened this morning, after a night's voyage upon the still pacific Mediterranean, by a voice outside my cabin, vibrant with emotion, hardly repressed:

"Have we reached Joppa?"

"In an hour, madame!"

Peeping beyond the edge of my door, I saw a woman gazing through an open port-hole with an expression that sent me to my own window. The east was flushed and golden in welcome to the coming sun; the sea dimpled with smiles; at the meeting of sky and water lay a dark irregular line of hills. It was my first glimpse of the Holy Land—and beautiful exceedingly! Yet, presently, I went back for another glance at my neighbor's face. Unconscious of possible scrutiny, her eyes were still fixed upon the horizon-line, and soul and thought went with them. She may have been fifty years of age, she was thin in face and figure, and plain-featured. She may have been a Yankee school-mistress or an English ex-governess. Something in her air forbade the supposition that she was illiterate or underbred; under the rushing association of the scene, the commonplace visage was glorified. In spirit she was on her knees before the altar of the hills, behind which gleamed the flame of a new day. I knew as surely as if she had broken into a Magnificat, that, like myself, she had longed through years for this hour; that nothing in human language could voice aright the emotions swelling her heart to actual pain.

Whatever may be the degree of disillusionment which many well-meaning people predict for us in our pilgrimage, I shall always be thankful that I met in that still hour, in spirit, the simple devoutness I read in the face of a stranger whose name I shall never know, and who will never suspect my reverent sympathy with her mood.

CHAPTER II.

IN BEIRÛT AND PARADISE.

"YOU have made a paradise here!" said a visitor to the owner of a city garden.

"Humph!" looking sourly at the dingy, ill-built cross streets by which his home and grounds were surrounded. "But you see I have to go through the other place to get to my Paradise!"

The growl recurred to me as we climbed the stone steps leading from the boat that had brought us to the Beirût landing. The yellow walls, red roofs and, what were at that distance, hanging gardens, of the town sloped upward to the brow of a hill that is not shamed by Mount Lebanon facing it across the narrowed sea. Tall palms showed above masses of olive, fig and mulberry trees and flowering vines.

At the head of the flight of stairs rocked and yelled a crowd in motley array. Had time and quiet been vouchsafed, we could have identified Parthians, and Medes, and Elamites, and the dwellers in Mesopotamia, Phrygia and Pamphylia and in Egypt. Strangers of Rome, Jews and proselytes assuredly had their representative waves in the surf of humanity, and Cretes and Arabians were loudest of lung, most violent of gesticulation. We were upon the topmost step when a big fellow in a black gown and a white cap reeled almost to the lower from a push dealt by a smaller man in custom-house livery. Before our turn came to stand in the official presence, two others were thrust from a door leading to an inner room. If there were one hundred men there, eighty-eight were vociferating in half-a-dozen different dialects, with apparently eighty-eight separate and dire grievances.

David Jamal—the impassively courteous dragoman—looked twice over his shoulder to say, "Keep close to me!" and in the lee of his broad back we waited until he presented our passports, with a grave bow, to the aforesaid functionary. He was not a large functionary, as I have said; he was young, and nature, in bestowing upon him a round face and fresh-colored cheeks, had not intended him to look fierce. He achieved a tolerable imitation of ferocity, as, jerking open the folds of the documents, and glaring from them to us, he demanded in English:

"Why didn't they have them *viséd* by the Turkish consul in America?"

We said nothing, but we were not abashed. David was running the affair; we had a serene sense of being only passengers. Had we taken part in the conversation, it would have been to ask why in the name of reason and the law of nations, we should have the papers *viséd* by the Turkish, any more than by the English, German, French, Egyptian, Austrian, Grecian and Italian consuls. As an illustration of the Turkish principles that might makes right and occupation signifies despotic power, our officer sat down at a table, after three minutes of

VIEW FROM MY WINDOW IN BEIRÛT.

heathenish raging, and executed a half-score of scratches in the corner of each passport, for which we paid six dollars and a half.

"Now, be off with you!" was the next mandate, and, still in David's wake, we passed into an inner and—incredible as it would have seemed a moment ago—a noisier court, where our luggage was examined. Men in braided jackets and red fezes, with swords at their sides, fell upon our respectable trunks, satchels and shawl-straps, as tigers upon sheep, tore out books, clothing and shoes, opened boxes and portfolios, ran rude hands into the depths and under the sides of the disordered mass. A wild exclamation burst from two who had in hand a shawl-strap containing a sea-rug and something hard and square. In a twinkling, buckles were loosened, the suspicious folds undone, and Alcides's kodak lay revealed to the majesty of the Ottoman government. The wolfish gleam in eyes that beheld the smiles we did not try to repress would have meant bastinado and bow-string a century-and-a-half ago.

While the examination was going on, I had retired to a bench behind a sort of counter, and, always serene in the persuasion that ours was a secondary interest in the scene so long as our dragoman stood dignified and unbending, in the fore-front of the battle, entertained myself with watching our fellow-victims.

THE "ECCE HOMO" ARCH, JERUSALEM.

One old Greek, with grizzled beard and travel-stained robe, had no less than six bags of divers materials, all sewed up with strong twine.

He actually threw himself upon them, as an officer whipped out a knife and began cutting the stitches.

A second official flew to his comrade's aid and tore the lean arms from their hold.

After that, they left not one stitch upon another to tell the tale.

Out tumbled upon the floor what looked more like the contents of a rag-bag than any possible valuables.

The officers kicked them apart and over, and leaving them where they lay, turned to the wretched owner, who was now blubbering like a seven-year-old baby, and dived into his pockets. From one they drew a filthy calico bag, containing what from the shape and size might have been marbles, and, after shaking it before his eyes, transferred it to the pocket of one of his tormentors. The old Greek sat like Marius in the middle of ruin, watering the dusty floor with his tears, when, our own belongings having been restored to the trunks and bags by David and his men, we left the screeching babel behind.

Nobody can form an approximate idea of the tumult accompanying what I have but faintly described, unless he has with his own ears taken in as much of it as the excruciated tympanum can bear. I had lain upon the lounge in my room in the Hotel d'Orient—a chamber with a ceiling fully twenty feet high, looking across the blue waters to the Lebanon range for an hour before the horrid din ceased to vibrate upon the nerves of hearing. Then—a card was brought to me, accompanied by a gift of fruit and flowers, and the gates of Paradise began to swing ajar.

They were wide open as we drove, later in the day, to the higher part of the town where are situated the buildings of the Syrian Protestant College—that splendid monument of Christian genius and faith in those who founded and have conducted it, and of the Christian liberality of large-minded men at home. As we cleared the lower streets, vegetation became more abundant and of tropical luxuriance. The yellow stone walls on each hand were overtopped by masses of verdure; passion flowers showed darkly-purple among the five-fingered leaves of vines running over verandas and house-fronts and along the rough walls.

Intersecting lanes were lined with tall cacti, the fleshy leaves as large as a man's hand, the stalks near the ground larger than a man's arm. Oleanders, pink and white, blossomed upon trees fifteen and twenty feet high; the regal "poinsettia," cultivated in American conservatories, spread broad disks of scarlet upon shrubs six feet in height; orange trees hung full of fruit, and under those laden with the dwarf variety of the same—known in our country as "tangerines,"—here and in Southern Europe as "mandarins"—the sweet aromatic globes lay as plentifully as windfall apples in a New England orchard. But the roses! the roses! growing rankly in every garden, and requiring little attention except from the pruning-knife—Marechal Neils, La France, tea-roses, yellow, white, crimson and pink—of which one can buy a half-bushel of buds and blossoms for fifteen cents--filling street and lanes with perfumes—how can white paper and black ink convey to my far-away audience an adequate idea of the wealth and beauty and sweetness in a Syrian rose-garden? Now and then, a whiff of subtler fragrance called our eyes to a mantle of jasmine draping a wall, and as our carriage drew up in front of the house built by a wealthy American for the President of the College, Dr. Bliss, whose name and fame are dear to all conversant with the story of Syrian missions, our wheels brushed against a hedge of rose-geranium.

Such were the external phases of our paradise on that late autumnal afternoon. Lovelier and far dearer was that unexpected welcome that awaited us from those who ceased to be strangers from the moment our hands met in greeting. This is not the place, nor is mine the province, to give a statistical history of the glorious educational institution—the light-house set on a hill—through which we were courteously conducted. We attended afternoon prayers in the fine chapel, the gift of another of our Christian countrymen. The services were all in Arabic—the selection of Scripture, the hymn, set to an Arabic air, and the prayer by Dr. Post—the distinguished head of the medical department of what is, to all intents and purposes, a university. The students are chiefly Syrians, many of them belonging to Moslem families, but there are Egyptians among them, and here and there an uncovered head bespoke Greek parentage or birth.

SCENE IN BEIRÛT.

Then we had what our English cousins know as "a cup of afternoon tea," in the cosy parlor of the President's house. My cup was the more delicious because passed to me by a bright-faced, sweet-voiced girl—a missionary in the third generation—to whom Arabic is as familiar as English. The tea-drinking was a prelude to a Thanksgiving gathering held two evenings later in the house of Professor Porter. Reckoning backward, as befitted our eastern pilgrimage, we computed that at the very hour in which we were thinking of and praying for "home and native land," hundreds of thousands of happy families were discussing Thanksgiving dinners. A sweet savor of turkey and pumpkin pies floated into the imagination, and as we sang, with a will, "My Country, 'tis of Thee," we could almost catch the echoes left in the welkin by the millions of voices that had shouted it that day. We had much and delightful social communing when the more formal exercises of the evening were over. To us, the newly arrived, the cordial hospitality then and there received, the pleasant ring of American voices, the genial warmth of inquiries as to our plans, and offers of co-operation in our work, are among the things one dare not trust oneself to commit to paper. As I write, the sunset is bathing the Lebanon Mountains with pink. The Lebanon, where Hiram's men relieved Solomon's in regular turn in lifting axes upon thick trees. The floats which were to convey the felled cedars to Joppa were moored over there in the bay. The wash of the waves mingles with the tinkle of the donkey-bells from the street below, that reminds one oddly of sleighing season.

"A BREATHING SPELL."

Circumstances have compelled us to tarry for a few days in this land of Beulah. The sea lies behind us, the hills and desert are before us, the breathing spell is welcome. Such wealth of kindness, such Christianly-affectionate treatment as has been ours from the hour we cleared the purgatory of the lower town, add sensibly to one's wealth of heart and memory. The recollection will sound through the years to come like the soothing murmur of the Mediterranean upon the Beirût beach; lie upon "mountain-ranges overpast" of experience, as the sunset flush upon Lebanon.

"Everything is ready for to-morrow, David?" as the majestic figure bows in the doorway.

"Everything, madame; my men, implements and animals."

"To-morrow, then, begins our real work?"

David bows again, his hand on his heart.

"Not of myself, madame, but with God's help, I hope to see you safely to your journey's end!"

CHAPTER III.

MARTHA OF LEBANON.

SHE looks up from her washing as we near her home. Her laundry is roofed by a friendly fig-tree; one side is protected from the wind by a wall of loosely-laid stones, picked up in the adjacent fields. Her house joins this at a right angle. Upon her right hand is a confused heap of baskets, water-pots, sticks and straw. A chicken roosts upon the topmost basket and will probably sleep in it to-night. A shapeless garment that may be Martha's own "izzar," or her husband's trousers, is spread upon the heap to dry.

The entire furniture of the laundry is in full sight, when we have added to it her "set tub." I need not remind American housewives of the insistence of Irish help upon this "convaynience" of every well-ordered household. Martha's tub is a great metal bowl, and it is set between her knees, her seat being the earth from which she—and we—sprang. One daughter sits with folded hands beside her; another is lazily stirring lentil pottage in the pot used awhile ago for heating the water required for her mother's morning task. For Martha, be it understood, is as sensible of the desirableness of having both "hot and cold water" when she is about her washing, as Bridget-of-the-bog. She makes a little go very far, however, and the fire of sticks over which it was heated having burned itself out into a bed of clear coals, she utilizes it and the kettle in the preparation of the family supper.

"SHE LOOKS UP FROM HER WASHING."

The family wash is not large. When she has rubbed, rinsed, and wrung a couple of white izzars for herself, a cotton tunic or shirt for her lord, and two or three sheets for the baby, the bulk of the work is disposed of. Those who groan over the "heavy wash of a-Monday" in our more (or less) favored land, can form an idea of the diminution of toil that would be brought about by the absence from the weekly tale of articles, of towels, table-cloths, sheets, napkins, stockings, handkerchiefs and underclothing of all kinds. Martha wears neither shoes nor stockings, nor does any member of her flock. Each is habited in at least one garment, for she is a decent body, and scorns the shiftless neighbors who suffer their boys to run abroad with no covering on their supple bodies. When the solitary frock or tunic or "combination suit" of blouse and trousers drops quite to pieces, a new one is substituted, and the remnants of the former are worked up in some manner. By the time that the washing is spread upon the stones and underbrush to dry, it is time the bread, set to rise this morning in a red earthen pan, should be kneaded. The kitchen is on the other side of the house. Martha has ideas of the fitness of things, and does not "clutter," as the manner of some is. She has all of out-of-doors to work in and uses what she needs. A cloth is laid under the low table which is sprinkled thickly with flour before the risen dough is cast upon it. What falls from the board during the kneading will thus be saved. With rapid touches she brings the yielding lump into a ball, and tosses it from one hand to the other with incredible swiftness. It grows in size and lightness before your eyes, and when she adjudges it to be light and large enough, is turned to the board and patted into a sheet hardly thicker than stout writing-paper. A round, slightly convex piece of iron, not unlike a shield in form and size, has been heating over coals glowing within a sort of semi-circular fender of red clay. The bread-sheet is laid upon the shield and cooks quickly. As each relay is browned, it is transferred to a wicker tray to cool, and another takes its place. These cakes have a top and bottom crust with a void space between them. They are usually sour. Indeed, most of the bread I have eaten, or tried to eat, in Syria, is sour, and my soul loathes it. Martha's cakes are likewise tough, shortening being an unknown quantity in the manufacture.

The family, as a rule, meet at but one meal in the day, and that is supper. Unless it rains, they will partake of this under the fig-tree, and chickens and dogs will be in readiness to devour such scraps as may be flung to them. Less careful providers do not get up a hot supper every day of the week, contenting themselves with preparing a great pot of stew of some description on alternate days, and eating what is left over on the morrow. Martha looks well to the ways of her household. Her husband earns a matter of thirty-two cents a day—which goes as far as a dollar-and-a-quarter would in the United States—and deserves a comfortable, orderly meal when he comes home in the evening. To-night, it will be "that same red pottage" we have seen the daughter stirring in the pot. We would call the lentils composing it brown; Martha knows them as red, and the dish she evolves from them is, we believe, identical with that bought by hungry Esau from his "smart" brother. The dried lentils are soaked, then boiled and drained, and left to cool and stiffen for awhile. Half-an-hour or so before supper, they will be thrown into boiling oil or fat of some kind, and heated thoroughly,

DRUSES OF MT. CARMEL AT MEAT.

then seasoned and poured into a large crockery or wooden bowl. If she would have it especially savory, she fries an onion in the fat before putting in the lentils.

"You will observe," says our interpreter, "that this is not porridge, nor yet what the French cooks set before us as 'potage,' or a 'puree.' It is a dish of pottage."

It gives forth a goodly smell as it reeks in kettle and bowl. I have eaten it with hearty relish more than once in hotel and private house, and never without

"HER FLOCK."

a sorrowful thought of the hungry hunter who made an unlucky meal of "bread and pottage of lentils."

He doubtless ate it as Martha's spouse will eat of this, from a bowl set flat upon the earth. The family gather about it, cross-legged, or, as one traveler puts it—"sitting upon nothing, as only Orientals can sit." Each tears off a bit of leathery crust from a round cake of bread, dips it deftly into the pottage, securing enough to envelope in the folded scrap of crust, which is thus conveyed to the mouth. As a fresh bit of bread is used each time the "sop" is dipped, the method is not unclean—provided always that the eater's fingers are clean. In view of the scarcity of water in the land, and the absenee of towels, not to mention the various uses to which bowls are put, we do not inquire wisely when we give this point too much consideration.

To-morrow, Martha may serve a stew of potatoes, flavored with onions and unctuous with grease. After the custom of the peasantry of all countries, the lower classes of Northern and Southern Syria are inordinately fond of fat and sweets. Potatoes, although introduced into Syria by missionaries, less than a century ago, have taken kindly to the soil and are a favorite dish. Or, she may pull up a cabbage from the plot of garden behind her hut, or buy it from a neighbor. Chopped cabbage, rice and onions, shining with fat, are not to be despised. There is also what we classify as a species of vegetable marrow, but shaped more like a cucumber, that, when shred into the pot, reminds us of the great vessel Elisha ordered his servant to "set on and seethe the pottage for the sons of the prophets."

BREAD-MAKING.

"And one went out into the field to gather herbs, and found a wild vine, and gathered thereof wild gourds his lapful and shred them into the pot of pottage; for they knew them not."

Rice is cooked far better by Martha than by a majority of professional cooks with us. Every grain stands up for itself, and is further encouraged to independence by a coating of olive oil or some kind of gravy. Seasoned with onion, and made more pleasing to the eye by bits of minced tomato, it is palatable and nourishing.

The hostess stands aside, and bows, raising her hand to her forehead, then laying it upon her heart, inviting us to pass into her dwelling before her. The outer walls of heavy stones are laid in sun-dried clay, with which the inner walls are coated. The flat roof is supported by rough boards laid from side to side. Upon these is a layer of sticks and straw, and over these earth is spread to the depth of eight or ten inches, and beaten or rolled flat. As the winter rains come on, the covering of the house must be frequently—sometimes daily—inspected. A crack soon widens into a fissure.

"By slothfulness the roof sinketh in, and through idleness of the hands, the house leaketh."

The continual dropping in a rainy day is of muddy water, and the household is fortunate if the whole superstructure of mud, twigs and boards does not give way and overwhelm them, or drive them out into the windy storm and tempest.

Our Martha is notable in her generation, and evidently expects us to admire the proofs of the quality offered by the interior of her abode. To appreciate her ingenuity, I must remark that the flooring of clay mixed with chopped straw or stubble is kneaded and spread by the women themselves, and that they keep the inner walls in repair by daubing them with the same mixture. When the floor is badly broken, or very foul from use, the house-wife cleans house by kneading a fresh supply of this untempered mortar and besmearing it anew. Martha has invented a process of combining her mortar with gum gathered laboriously from mountain-trees, which imparts a gloss to walls and floor. She has likewise smoothed the cement with a board instead of contenting herself with patting it level with hands and feet. The only attempt at decoration we have seen in the houses of the Eastern peasants is prominent in the ten-by-ten room in which we now stand. It is Martha's masterpiece and must be duly admired if one desires to secure a reputation for taste in the fine arts. Right in the middle of the room (which is the house) is a rude pillar, about as thick and high as a five o'clock tea-table, constructed of red clay hardened by gum, and really resembling red stone.

Besides the door there are two small windows upon opposite sides of the apartment. The dead-wall facing us as we enter is the family store-room. The place of honor is given to the barrel of family flour (meal). Upon the top is the kneading-trough, or pan, in which the dough is mixed and set to rise. A hole near the bottom of the barrel lets out the meal when it is needed for use, and is then plugged up with a rag. The sight corrects our pre-conceived ideas of the widow of Sarepta's mode of getting the day's supply from her exhaustless barrel. We who have pictured her as leaning far over the edge, scraping up the flour from the echoing bottom, have lost the beauty and the significance of the miracle. The daily dole came from above, as will be seen, and she had no means of estimating how much or how little was left in the barrel, as she drove the plug back into place.

The cruse of oil, by a pleasant coincidence, not of our contrivance, is near by on the floor, full-bodied and narrow-necked. If faithless, the widow might peer, but unsuccessfully, to guess how long she might depend upon the contents remaining in the bottom.

I cannot leave the widow of Sarepta without relating a little incident that came to me the other day. A party of travelers and missionaries visiting Sarepta from Sidon on Thanksgiving day saw, just without what was once the gate of the city, "a woman gathering of sticks" to make a fire.

The commonest event of the lowliest life in this country, so strange and yet so familiar to the Bible student, is an expressive commentary upon the Scriptures.

As when, before coming indoors, Martha, perceiving that the fire of coals under the lentil-pot was nearly out, hastily rolled a bunch of dried grass into a wisp, thrust it under the kettle, and as it smoked, persuasively blew it into a flame.

"If God so clothe the grass of the fields, which to-day is, and to-morrow is cast into the oven."

"Smoking flax shall he not quench."

But to our store-room ;—Strings of onions and peppers hang from the ceiling; smaller barrels of rice and other commodities are in line with that holding the meal; jars, pots and kettles are upon these; a big pestle is suspended alongside of a miscellany of baskets, cloths and porringers, each of which has value in Martha's sight. Her cousin of Bethany would hasten to unroll a mat or rug for us to stand upon. It is not easy or convenient for country Martha to get woven matting, and rugs are a city luxury. We can see for ourselves that the broom of twigs set in house-wifely order, with the brush end uppermost, against the wall, has been in diligent use to-day. The floor, if hard, is smooth and clean, and her Bethany kinswoman could not offer with more cordial hospitality the cushions now brought forward for our accommodation.

We dispose ourselves upon the cushions, my guide and instructor—Syrian by birth, American by education, pure womanly by nature, and earnest Christian by grace—with the ease acquired by years of practice, I more comfortably than before I began the round of visits that has engrossed me of late. The cushions are round, long, stuffed with hay, and covered with Turkey red cotton. While the gentle voice that lends itself musically even to Arabic goes on with friendly discourse with the mistress of the home, I have opportunity—given me purposely—to inventory the furniture.

I amuse myself with imaginations of how the wife of a New Jersey or Illinois day-laborer in steady employment would behave if introduced to this place as her abiding-place for the rest of her natural life. On two sides of the room runs a wide shelf of unplaned boards, elevated upon sticks about a foot from the earth which is the floor. On this is piled the family bedding in the shape of "pallets"—two breadths of cotton cloth run together at the edges, and wadded with hay, or cotton, or cloth, or rags, to the thickness of an inch or so. Upon one of the quilts the would-be sleeper will lie to-night when it is unfolded over the floor, and with another he or she will be covered, the nights being chilly at this season. Father, mother and five children will rest upon these pallets, laid as closely together as they can be put, and having known no other beds, their rest will be sweet in spite of the unyielding floor, the low ceiling and the stuffiness arising from seven pairs of lungs and seven unwashen bodies. Windows and doors will be closed, for the night-air is believed to be unwholesome. Martha would also ask in grave wonderment when she hears of the daily baths of her pale-faced visitors, if they "are not afraid of making themselves ill by so much washing?" She is neat and notable according to her lights, but illumination has never struck athwart the, to us, vital question of much soap and more water.

Before judging her, I recall the reply made to me once by a New York woman who has thrown talents, energy—all that make up and abound in her grand personality—into the work of elevating working-women to a higher plane of thought and living.

"They might at least make themselves 'skin-clean,'" I heard said, "even if they have not time for keeping their poor rooms tidy. Bathing costs little."

"You might change your mind if you had to perform your ablutions in a quart of water, brought by hand—and a very tired hand!—up four flights of rickety stairs," retorted the practical philanthropist.

"BROUGHT UPON HER SHOULDER."

Every drop of the water used in Martha's household is brought from the fountain in the village at the foot of the hill, and brought upon her shoulder in a huge jar of baked clay. The recreation she derives from gossip, friendly or mischievous, with other burden-bearers who there do congregate, hardly offsets the bodily fatigue of trudging up the steep and stony path. Her husband stays his labor in the threshing-floor and, leaning upon the "fan" in his hand, throws her a kindly word as she passes, the drip from the water-pot mingling with the sweat of her face. He is a model spouse, as Moslem spouses go, but he would no sooner think of offering to run up the hill with her load than Eleazar of Damascus thought of forestalling Rebekah in her hospitable task of "watering all his camels," after she had let down her pitcher from her shoulder and given him to drink.

As for her boys, she spoils them from the beginning, teaching them by object-lessons, if not by precept, the inferiority of her sex to theirs. Sooner than degrade them to the position of drawers of water, she would fetch and carry every hour of every day of all the days of her life. In due course of time they will marry and each bring his child-wife home for at least one year to be drilled in housework by her mother-in-law. Martha will get her "innings" then, and lives in hope of this consummation.

To-morrow morning, she will hustle the laggards out of bed and out-of-doors, that she may roll up the beds and pack them upon the shelf. The cushions keep them company except when a visitor drops in, and besides these, there is absolutely no furniture in the house—neither table nor chair, nor a stool, nor anything bearing the faintest resemblance to a chest of drawers. The family wardrobe is upon their backs. If there are extra garments they are in sundry uncouth bundles and a couple of rough boxes tucked under a broad shelf and among the bedding. Her pots and kettles, jars and bowls, compose the "plenishing" of the home.

"But when it rains," I ask, "where is the cooking done then?"

Martha brings out a misshapen utensil of clay baked in the coals, which is enough like the braziers in use by tinkers and solderers to make it recognizable as a receptacle for fire. She explains how this is set upon the pillar in the middle of the floor, a fire built within it, and what could be more satisfactory?

"Where does the smoke go?"

A gesture replies—"Anywhere!"

"And the children, who cannot run out of doors in the rain?"

She admits that they are troublesome in such circumstances, but it is the will of God that the rain falls, and she must make the best of it.

I think of the five restless little beings, wild as hawks and wayward as the wind, and the curling smoke under the low rafters, and the dampened roof making dark the interior, and find the only light in the picture in the reflection that real winter and Syrian rains—violent deluges as I know them to be—last, as a rule, but two months a year.

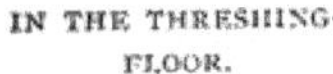

IN THE THRESHING-FLOOR.

She will have few of the morning tasks to perform that render the day-dawning a bugbear to Occidental house-wives. When the mats are stowed away, a cup of coffee made for her husband and the children served with a cake of tough bread apiece, to be devoured when and where they please, the day may be said to be fairly upon its feet. Her husband has carried his luncheon to the field with him; the children will be nibbling all day long, but there will be no noon-day meal spread or cooked. It is surprising, by the way, what an amount of various trash those childish stomachs try to dispose of during the day. Besides the leathery cakes that are the chief of their diet, there are radishes in abundance—big, juicy roots—to be had for the pulling out of the ground, and raw turnips, and onions, and salads, and such homely sweets as the mother compounds. They run about with slices of cold, boiled beets in grimy fingers, devouring them as greedily as if they were candy, and chew and suck bits of purple sugar-cane, and there is a tall jar of "dips,"—i. e., grape-juice, boiled down to the consistency of treacle—in the corner of the house, to which resort is made when the mother's back is turned.

Candidly, we cannot see what Martha can find to keep her as busy all day long as she assures us she is. That she believes herself overworked is evinced by the worried plaits between her eyes, as deep as the furrows plowed by ceaseless toil and care in the forehead of a New England farmeress. Yet this is the easy season among the dwellers upon Lebanon slopes, for the silk-worm work is over for the year, and the eggs that are to be hatched next spring are simply done up in bags and hung up in the church over yonder—less out of custom or superstition than because the eggs must be kept at an even temperature, and out of the children's way. For these ends there is no safer place than the church. All the neighbors thus utilize the sacred place, and the priest raises no objection. Martha does not know how she could pay the rent of her house and keep the children clothed, but for the blessed worms. From the beginning of the season, when the eggs are disposed upon boards and hurdles in the hut, to the day when the last cocoon is completed, the family dwell in booths out-of-doors, leaving the quiet dwelling to the voracious feeders upon the mulberry leaves gathered fresh daily for them. They eat the whole of the first crop; a second puts forth for the sheep.

"IT IS QUIESCENT UNDER TREATMENT."

This last word is in the singular number in more than one sense of the adjective, as our illustration will show. The oldest girl of the humble home has her hands full for some hours of each day in collecting food and literally stuffing it down the throat of the animal. It is tied, and could not get away if it would. It soon ceases to frisk and tug at the cord, and is as quiescent under treatment as the Strasbourg goose whose feet are fast to a board, and whose liver distends abnormally in the attempt to digest the matter with which it is surfeited. The sheep belongs to a breed that has very broad tails. As he becomes first plump, then fat, then so unwieldy that he can hardly breathe, and does not offer to move, the tail grows to an enormous size, often weighing from thirty to forty pounds. To the day of his death, which must be a relief to the ponderous body, the stuffing is kept up, and he must have consumed many times his weight in mulberry leaves. The horrified beholder marvels within his disgusted soul if the creature could not be spun into silk as well as the worms, which are monsters of gluttonness and corpulency.

Of the skin, with the fleece on, Martha has made, about every fifth year, a coat for her husband, who sometimes has work further up the mountain in winter. If this be not worn out, she finds a ready market for wool and hide. Every morsel of the precious fat is treasured, "tried out," and clarified, then poured into jars to harden, and used during the winter as butter. Few cows are kept by the peasants in the country, none in towns. When milk must be had, they get it from goats, or buy from better-off neighbors. All classes are fond of what is called "lebben," which is nothing more or less than loppered milk, "turned" into firm blanc-mange-like consistency by adding to sour milk a "lebben" left over from the previous day. This makes the delicacy more sharply acid than our loppered milk, or "bonny clabber," as the Southerners name it. I have seen "lebben" passed with a dish of rice and meat at a hotel table, and eaten from the same plate as we would a vegetable or sauce.

"HER TURF-ROOFED HUT."

The bones and meat of the valuable sheep are boiled until the flesh slips off easily, when it is cut into small pieces, seasoned well and packed into jars. Melted suet is "flowed" over the surface, and the jars are set in a cool, dry place. Except when a chicken is killed upon some great holiday, the family have no other meat than this as long as it lasts.

Martha reckons fat of whatever kind as more valuable than lean meat, and, next to olive oil, the stored suet from her fatted sheep as most desirable. It is, really, sweeter and more delicate than could be imagined of "mutton-tallow," and should not be confounded with it.

I have rated the thirty-two cents a day which represent the earnings of the head of this Syrian house as equal to a dollar-and-a-quarter or a dollar-and-a-half in United States currency. I am told by trustworthy authorities that this is a fair statement in view of the wide disparity between the customs of the two countries, as well as between the needs of the poorer classes here and in America. Martha does not need to lay in coal or wood, or to keep up fires night and day. Charcoal is cheap; sticks may be had for the gathering, and in the lower countries dried camels'-manure is sold at a trifling cost in the markets for fuel. The boys need no flannels, the girls no shoes, no bonnets or caps, there are no Sunday clothes, no bedsteads, bureaux, washstands, chairs, or even tables.

Tea is an unknown luxury, and when she uses coffee, it is of the commonest sort. There is no striving to keep up appearances. So long as she lives as well as her fore-mothers and her acquaintances, and a little better than the majority of the last-named, she is content. Her world, to our eyes, may be typified by the compass of her turf-roofed hut, but it is all the world she knows, or is likely to know.

CHAPTER IV.

AN AFTERNOON CALL.

HAVING due regard to the proprieties it is, perhaps, well that I should not state in which of the larger cities we have visited in Palestine and northern Syria I availed myself of the invitation through a friend to pay my respects to the wife of a wealthy citizen. In our wanderings we have come upon Boston, New York and Chicago newspapers in such unlikely places that we have grown timid in the use of the names of persons and well-known localities. Some people on this side of the Mediterranean (and on the thither side of the Atlantic) like to be written up. The sight of one's name in print infuses a grateful glow through the moral and mental system, and titillates the self-love which is the one mighty common attribute of humankind from pole to pole. Others shrink honestly from printed publicity, as from the touch of red-hot iron, even if the notice be laudatory. Nobody—and this is a rule without exception—enjoys being made ridiculous in the sight of friends or strangers.

In view of these things, the sketch of this one of my visits, if I would report everything as it happened, must go forth undated as to time and place.

We were set down by our coachman at the mouth of a long, rather narrow passage or court, paved with rough stones, and neither light nor clean. In the United States, we should have considered it an unpromising entrance to a factory or warehouse. As the introduction to the abode of one who counts his wealth by millions of dollars, it was simply inexplicable to the mind of the average traveler from the sunset land. As strange seemed the four flights of stone stairs we climbed to reach the drawing-rooms. They were not even marble steps up which we toiled pantingly, but of the yellowish freestone in general use here for the better class of buildings. Staircases borrow new horrors (to those who have left youth and length of wind behind them) from the height of Oriental ceilings. It is not unusual to find these in private residences from twenty to twenty-five feet high, and each staircase conducting to the floor above is double and mercifully broken by a landing. Upon these landings we paused to gather breath and heart, until we beheld at the head of the eighth half-flight a maid in jacket and short skirt, a veil pinned over her black hair, awaiting us. Taking the right hand of each of us in turn with both of hers, and courtesying so deeply that one knee must have touched the floor, she raised the back of the hand, first to her lips, then to her

PEASANT MOSLEM MARRIED WOMEN OF PALESTINE.

(60) MINARET IN JERUSALEM.

forehead, lastly to her heart, and stepping back as she arose, motioned us to precede her into an open door. A few steps within the threshold of a spacious drawing-room we were met by the mistress of the house. She is a Circassian, married to a Turkish gentleman, and still bears traces of unusual beauty in her clearly-molded features, dark eyes and sweet smile. Her attire was a disappointment. I had expected Oriental magnificence and found Parisian simplicity in a pale-pink gown, trimmed modestly with native lace. Except that her wrists were loaded with bracelets and that the brooch at her throat was a superb emerald, the largest I ever saw, surrounded by diamonds, I should have sought vainly for tokens of her nationality and her husband's wealth. She greeted us cordially, shaking hands as an Englishwoman might, and led us through the outermost and largest room of the suite to a smaller, where she waved us to arm-chairs. Two little girls, eight and ten years of age, were their mother's aides in making us welcome. Each bent her pretty head to kiss our hands, and each, unbidden unless by the mother's eye, hastened to fetch stools for our feet before withdrawing into the background.

I could speak no Arabic: our hostess neither English nor French. We carried on a sort of three-cornered conversation that would have been droll enough to a listener, the trite nothings of polite society losing what little flavor they might have had at first hand (or mouth) in filtering from English into Arabic and back from Arabic to me in English.

My initial observation that the afternoon had become suddenly warm was made in French under the impression that the handsome Circassian understood that tongue, and at her inquiring look, had to be "done" into English that the interpreter might get hold of it. It was irresistibly funny, but we three kept straight faces and proceeded to pelt one another with cut-and-dried figures of speech, the little daughters regarding us with wide, grave eyes, and a maid, a straight, dark-eyed woman, a striking figure, standing in the arched doorway of the third drawing-room, never removed her gaze from the hands of her mistress. She had slipped off her sandals upon entering the parlor, and stood now in her white-stockinged feet upon the carpet.

This carpet was another and a sharp disappointment. It was costly in material, but for pattern might have been selected by a rich vulgarian in Cincinnati or Denver. With the vision before me of certain Persian and Turkish rugs I had seen that day in a city bazaar, the soft harmony of whose colors was a dream of artistic perfection, I resented the glaring flowers ramping over about four hundred yards of white ground. Chairs and sofas from a Parisian upholsterer were pushed stiffly against the walls; satin curtains from the same establishment hung at the dozen windows. The bare walls were the only evidence of Moslem occupation of the vast rooms.

I was beginning to weary of tasteless splendor that had in it hardly a touch of the Oriental element, when another maid, putting off her sandals at the entrance, ushered in two more visitors. They were introduced to us in due form as the wife and mother of the highest Turkish official in the city, and were evidently personages of importance. It presently transpired that they were to remain to dinner. That meal would be served at six o'clock, and it was now a little after four. Three maids removed the new visitors' hoods and wraps, and the wife of the Pasha accepted an arm-chair. Not so with her mother-in-law. Parisian innovations might catch the fancy of the younger generation, and turn aside the very elect from time-honored customs. She would none of them. A couple of maids hastened under the mistress's direction to heap in one corner four red satin cushions, and, when the old woman had crossed her legs upon them, to place other cushions at her back. Thus established, she sighed satisfiedly and smiled around the room. She supplied the "Oriental element" for which I had longed!

Answering the smile and nod bestowed upon me when she learned that I had made a three weeks' journey to see her, I took a mental photograph of her as Moslem and as mother-in-law. The East is the paradise of mothers-in-law, as I shall further demonstrate some day, and this particular specimen of a well-abused class magnified her office. It was plain that she considered herself the most important personage in the room. Her daughter-in-law was languid and looked sickly. Her sallow complexion showed to disadvantage against a Parisian gown of tan-colored stuff, with a vest of creamy silk; her heavy bracelets hung upon her lean wrists like handcuffs; her head was bare, and her hair negligently arranged. After her mother-in-law had interrupted her several times in the middle of a remark indolently uttered in the Turkish dialect, she became silent, leaned back in her arm-chair and lighted a cigarette. A silver stand of cigarettes and matches was brought in just before her arrival.

The "Oriental element" was clad in yellow silk trousers, white silk stockings and yellow slippers, as her attitude allowed us ample opportunities for observing. Above these was a long shapeless sacque of fawn-colored silk, quilted in perpendicular rows. If she had selected color and pattern with an eye to heightening her native homeliness, her success was perfect. The sacque was large in the neck and revealed the dingy wrinkles of the throat. A big diamond on one hand and a ruby upon the other called attention to the thick, yellowish fingers, when she clasped her hands below one knee and swayed gently back and forth. Her head was bound about with a white turban, or loose cap, below which escaped stray gray locks; she had but three teeth, and her lower lip hung loosely. I could compare her to nothing but the hag of childish nightmares and the wicked fairy of folk-lore. Yet she laughed benevolently when the pretty little girls approached her and bowed to kiss her veinous hand, and patted them on the head, muttering what sounded like an incantation, and was probably a blessing.

Small gilded tables were now brought in, and upon them were set in array a Sevres tea-service, and silver baskets of biscuits and cakes. The hostess poured out the tea, through a silver tea-strainer—precisely as an American woman serves it on her "At Home" afternoon; the maids passed the refreshments to the guests. The "Oriental element" refused the foreign beverage with an imperious wave of the hand, and looked on rebukingly.

But the oddest performance was still in store for my unaccustomed eyes. As the bell in a neighboring tower struck five, the old lady scrambled to her feet like a cat, and without uttering a word, waddled into the third room of the suite. The black-eyed maid who stood, statue-like, in the doorway, followed her, and the hostess, with no sign of surprise, excused herself to us with a slight bend of the head, and went, as we supposed, to see what it all meant. In my Western ignorance, I supposed that the "Oriental element" meditated a nap upon the luxurious divan visible through the wide arch, and when the maid dragged out a rug from beneath the cushioned lounge, was puzzled by seeing her lay it in the exact middle of the apartment. The hostess now produced from the drawer of a cabinet a long white scarf of some thin tissue, and offered it to the "Oriental element." Stepping upon the oblong bit of carpeting which I now saw was what buyers of eastern stuffs know as a "prayer-rug," the crone wrapped the scarf over her head and across her forehead and chin, after the manner of a veil, and folding her hands, first upon her chest, then upon her forehead, raised them and her eyes toward the ceiling.

The maid resumed her station in the doorway, her dark face as immobile as the Egyptian Sphinx, her eyes again directed to the hands of her mistress, who came back to us, saying tranquilly in Arabic:

"Madame wishes to pray!"

I bent my head involuntarily, feeling and courtesy dictating some token of respect to an act of devotion, but the daughter-in-law took a fresh cigarette and a handful of sweets from a basket near by, and with the air of one temporarily relieved from an irksome presence, began a lively chat with the lady of the house, into which my friend and interpreter was speedily drawn. While the talk proceeded, I watched furtively, but closely, the proceedings in the adjoining room. It was one of the five times a day in which the devout Moslem must pray with his face toward Mecca. I have been informed since that it is unusual to see a woman Moslem at her devotions, although in this part of Syria the faithful of the other sex drop spade, trowel, paint-brush, chisel, awl, plow, even knife and fork, at the appointed season and prostrate themselves as commanded by the Koran. But this particular specimen of the "Oriental element," obviously differing from St. Paul as to the profitableness of bodily exercise, went through the various stages of her orisons with a swing and energy worthy of the most stalwart follower of the

Prophet. She bowed seven times toward the holy city, at an angle her apparent infirmity would seem to make impossible; she lifted eyes and clasped hands times without number, and finally went down upon her knees on the prayer-rug, and bumped her head hard upon the spot indicated by the pattern as the proper and edifying point where the faithful forehead should strike. She was still in this attitude, her covered head rising and falling with the regularity of an automatic toy, when we took our leave. The performance, so extraordinary to us, was of no moment whatever to her fellow-religionists. Their gossip went on more smoothly for her absence; their voices were not lowered by so much as a semi-tone. Unless deafened by religious ecstasy, she must have heard every word. The hostess offered no excuse or explanation. Attending us to the outer door of the drawing-room, she thanked us for the honor we had done her by calling, and hoped that we would repeat our visit, then went back to the Pasha-ess, who was helping herself to a third cigarette. The little daughters kissed our hands, and raised them to their foreheads, the three maids courtesied low to us from their several stations; our last glimpse of the "Oriental element" showed her still prostrate, and still smiting the prescribed spot of the prayer-rug with her forehead.

The fair Circassian is her husband's only wife, and they have, as we have seen, advanced ideas upon the subject of housekeeping and the entertainment of visitors. Their daughters have a governess—a Parisian—and the sons a tutor. But with the exception of her husband, not a man enters her presence. There was a queer sensation in sitting in social converse with three women whose lives were so rigorously divided from what, in our own land, brings variety and spirit into homes and society—the frank and gracious association of the sexes.

Woman may be said to supply the sugar-and-water at such entertainments, and some people, notably the French, enjoy "*eau sucre.*" The American palate—and constitution—give the preference to lemonade.

CHAPTER V.

A SYRIAN BABY.

THREE months before our call the news that a birth was impending drew about the one-roomed hut in which parents and three children already lived, a crowd of friendly neighbors. Outside, the men smoked and talked with the father to while away the period of suspense. Within, their wives thronged the chamber to suffocation, also smoking and talking, all at once. The mother lay upon a mat at the far side of the floor, with just enough room between her and the wall to allow the passage of the loudly-officious matrons who hovered about her.

Dr. William Gray Schauffler, the beloved physician to many natives in Beirût, once told me how he had charged into the midst of such a rabble, upon being called to take charge of "a dangerous case."

"There were fifty women in a twelve-by-twelve room," he said. "The air was stifling; the hubbub indescribable. I wasted no words. Two strong sentences in Arabic, with gestures to match, sent them scurrying to the door, and I could at least *see* the patient."

When the sex of our particular baby was reluctantly announced, there was a general falling away of solicitous friends. A dead silence followed the unwelcome phrase—accentuated, presently, by the groans of the nearest of feminine kindred and the sobs of the poor mother. The men emptied their pipes upon the ground and stalked off, mercifully forbearing to look at the father, disgraced by the appearance of still another daughter.

The women departed in like manner, without speaking to him, or to his wife. Often the husband approaches the mat upon which the unhappy woman is left alone with her new-born child, and scolds her vehemently for the disappointment she has caused him. Sometimes he actually strikes her in his fury. My guide told me of one instance that had come under her personal observation in which the mother had beaten her head and breast with her fists in a frenzied attempt to commit suicide. Life under the shame that had come to her was insupportable.

Had our baby been a boy, a turmoil of congratulations would have ensued upon the proclamation at the door of the hut. The happy father would have been embraced with tears of joy by his comrades, drums would have been beaten and trumpets blown, and such humble gifts as the poor can make to one another been sent in to the mother, already overwhelmed by the caresses and praises of her gossips. The little intruder upon a domestic circle and community thus organized had had few visitors and no presents when we made our call.

A SYRIAN MOTHER AND HER CHILD.

"I wish," said an indignant woman-missionary, "that not another girl would be born to you Syrians for half a century, that you might know the real value of woman in the world!"

The cradle of roughly-cut, unpainted wood stood in the middle of the floor. Upon slats nailed over the tall rockers was tacked a mattress stuffed with straw. Directly upon this, the child, clad in a single calico garment, was strapped by means of strips of cotton cloth, six inches in width, attached to the framework of the cradle. The only pillow was placed exactly beneath the shoulder-blades, the head sagging forlornly upon the lower level of the mattress. The baby's arms were laid close to its sides, its legs were stretched as straight as though the bed were its coffin; a sort of "duvet"—or quilt—was spread above it, and over all the straps were passed, smooth and level, under the slats and up again on the other side, to be tied with tape strings to a rude framework above the head of the tiny mummy. She could not stir finger or foot, or roll over by so much as a quarter-inch of space. When she cried, the mother knelt upon the floor and nursed her.

"She does not like it—no!" she answered, smiling at the query whether or not the baby relished the strapping process.

Nevertheless, whenever the small prisoner cries, it is assumed that she is hungry and whatever the mother is doing, she leaves everything to feed her. The custom of regular meal-time practiced by most intelligent mothers and nurses in America and England is considered barbarous.

"But it may be that the mothers in those countries do not pity the poor little things as we do?" said our hostess, tentative, but courteous, as she responded for the third time within half-an-hour to her nurseling's fretful appeal.

It stopped whining when she lifted it to display the construction of the cradle and the mysteries of the long strips of cloth, depositing the late occupant upon the mat beside her. The brown legs were kicked eagerly into the air, the arms tossed wildly in true baby-style, and the gurgling coo which is the natural language of infantile humanity the world around testified to its relief and pleasure. A four-year-old sister laughed gleefully in reply, and grabbing the baby, hitched it up to her hip in Syrian fashion, the tiny head wobbling frightfully.

"She will surely drop it!" gasped I, with a glance at the pitiless floor, a concrete of clay and stones.

The mother smiled; the interpreter reassured me.

"Oh! the older children carry the babies—when they are carried—from their birth, even when just able to toddle about themselves. She has probably had that little one in her arms four or five times a day, ever since it was born. The boys never touch them, of course. The girls are burden-bearers from the first."

The baby's head was already one-sided, the soft skull yielding readily to the pressure of the straw mattress. For twenty hours out of twenty-four it is bound down in the way I have described, and as the mother always kneels on the right-hand side of the cradle to nurse it, the head is habitually turned in that direction.

"There is hardly one well-shaped skull among the hundreds of Syrian boys and men in our preparatory school and college," said one of the Beirût professors to me. "They are all flattened at the back, or on the side, before the first year of infancy is over. They wear the fez indoors and out. When it is removed the deformity is seen."

THE GIRLS ARE BURDEN-BEARERS FROM THE FIRST.

If the effect of the much-dreaded evil eye be, as is supposed, a mysterious wasting away of flesh and loss of vigor, some former visitor may have exercised it upon another wee creature in a swinging cot. She was so fragile in frame, so ethereal in her pale prettiness, that I asked out of the fullness of common-sensible experience and observation upon what diet she was fed.

"She is not yet weaned," was the unexpected answer, "but she eats anything, potatoes, garlic, radishes, green almonds, cucumbers, bananas, cactus-figs" —the fruit of the wild prickly pear that flourishes rankly everywhere—"everything in fact, that people grown like. Why not? she has twelve teeth."

While talking, the mother—a Druse—squatted upon the bare floor, having laid cushions for us. She was comely with her black eyes, white teeth and ready smile. Her dress consisted of pink calico trousers, very wide and full, a short jacket of the same material, of a deeper red, and a veil of coarse white cotton, pinned tightly around her forehead and falling down her back. The sale of red calico of every conceivable shade must be immense in this country, almost as large as that of the blue cloth of which the men's jackets and trousers are made. While we chatted of the baby, a third was brought upon the scene, of course in the arms of an older sister. Few of the babies I have seen are plump, and fewer are robust with the rollicking elasticity of muscle and joints manifest in healthy American babies. Yet the much-decried "state of artificial civilization" is as far removed from them as if they were creatures of a different genus from our rosy, well-fed darlings. Those who advocate bringing up children in absolute obedience to natural laws and instincts would have few suggestions to offer to the Syrian mother, except in the matter of binding her baby in the cradle. As soon as he can creep, he is tossed upon the bosom of Mother Earth, and left to "hustle" for himself among his fellows. When drowsy, he crawls into a corner and goes to sleep like a little brown dog; if hungry, he eats whatever comes within reach of his dirty hands that commends itself to his judgment as possibly eatable. At night, he huddles down, in the one garment he has worn all day, close to brothers and sisters, to gain warmth through the sunless hours from their bodies.

When his mother is in a good humor she strokes and pats him and flings a fig or morsel of sweet cake to him; when she is busy, she kicks him out of her path; when angry—and this is often with the ignorant, untrained woman who was married at twelve years of age—she takes a stick to him and swears volubly, cursing the day in which he was born and invoking the vengeance of heaven upon his undutiful soul. The immature soul that, in this state of nature and natural development, gets even less washing than his body!

"How clean your baby looks!" said my guide to the mother of still another baby.

It was a boy, and almost the only infant we had seen whose skin was clear, and whose hair had the appearance of growing upon a healthy scalp. The proud parent showed her white teeth in a gleam of childish gratification.

"Yes! He should be clean. He had his first bath to-day. He is four months old, quite old enough to be taken to the baths."

Home washing is an unknown luxury. Four times in a year, parents and children treat themselves to a plunge at the public baths, where a cleansing can be had for an absurdly small sum. For three months thereafter, soap and towels, comb and brush—belonging to the "state of artificial civilization"—have no place in the unsophisticated household. While we still sat about the swinging cradle, the rush of naked feet was heard, and three little girls from five to ten years of age halted upon the threshold at sight of the clean new matting the mother had unrolled at our entrance, and, one after the other, stepped into an earthen bowl of dirty water standing outside, dabbled their toes in it, rubbed one bare foot hastily over the other and entered. They left wet tracks upon the uncovered strip of concrete flooring, and muddy marks upon the "company" matting, but complacent in the conviction that they had done all which decency and etiquette demanded, they tucked their dripping feet under them, and composed themselves to get their share of the enjoyment of our visit. Of the unmarried girl of Northern Syria and Palestine I will talk at some future day.

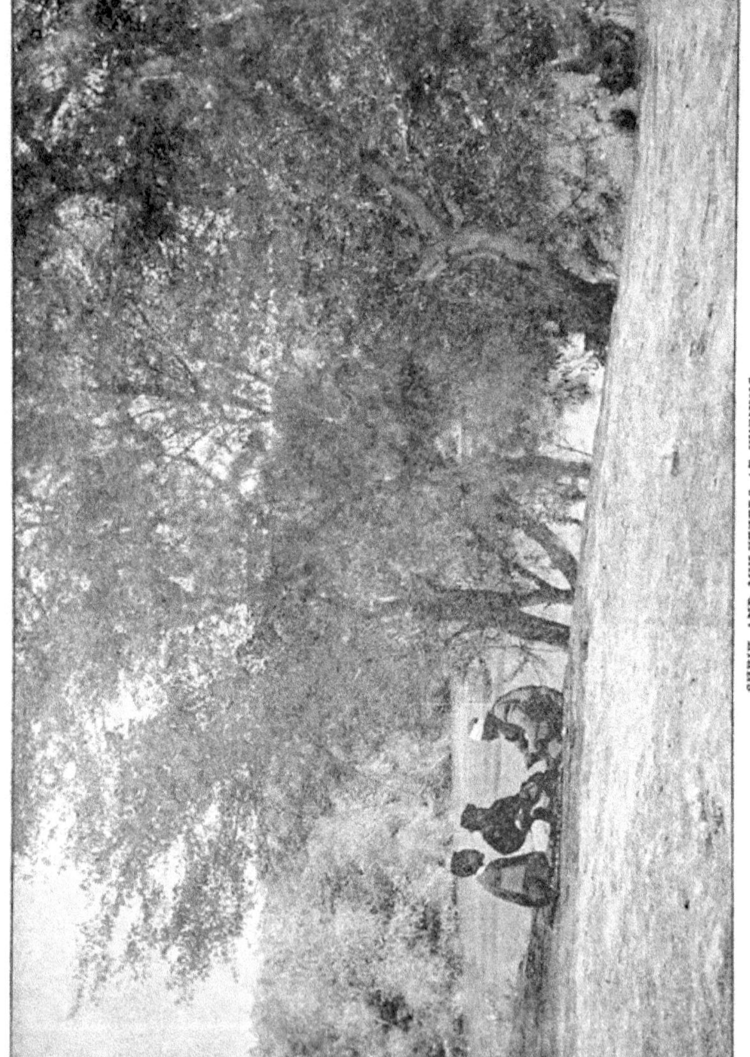

SHEIK AND MULETEERS AT EVENING.

CHAPTER VI.

IN DAVID'S CAMP.

"As when the weary traveler gains
  The height of some o'erlooking hill,
His heart revives, if 'cross the plains,
  He sees his home, though distant still."

I REPEAT the lines, wistfully, from a "height" so lofty and crowning a descent so steep that all have alighted from their horses to walk down. David Jamal has lifted his hand to point down the valley lying at the foot of the range:

"There is our camp!"

According to custom, the mule-train was sent on early this morning to make all ready for our coming. The little encampment is always a goodly sight. It was never more inviting than as we see it now, nestled among olive trees and backed by a green plain. We give hardly a glance to the range beyond, which must be passed to-morrow. Visions of precipitous defiles and sudden "drops" in the road, and hills full of rolling stones, to me most formidable of all, where the formula, "I believe I will walk up this!" excites the good-humored smiles of the escort—none of these things mar the pleasure which warms our hearts at the first glimpse of the white tents that, by now, mean home to the wanderers. We walk cheerily down the rugged slope, and more cheerily traverse the stony level at the bottom, a river in the rainy season, and now filled with pebbles and boulders brought down by the freshets; blithely we climb the knoll on which smiles our encampment.

David and the Bedouin sheik, who is our safe conduct through the country, have spurred forward, and the former stands ready to lift me down at the door of "the lady's tent." Within, water and towels await me, and a bed as comfortable as I have found in any hotel, should I care to rest until "afternoon tea" is served. By David's wise management, we are seldom, if ever, too late in arriving for this refreshment. By the time I have washed face and hands, and brushed my hair and garments, Imbarak, tall and serene, has set the tray in order, our camp-chairs beside it, and awaits further orders. The tea is swallowed and our hearts are revived into actual joyousness.

"The jolliest life in the world!" sighs Alcides, in blissful content. "Don't get out your note-book. It is enough to be alive on an evening like this—and here!"

We never weary of the incidents and scenes of camp-life and continually discern new and picturesque features in it. Opposite the door of my tent is the luncheon-booth, different in shape from the rest, with an awning above the entrance, never let down to close up the interior, except when rain overtakes us on the road, for it accompanies us everywhere, and is the only tent pitched at noon. Like the others, it is lined with a vari-colored fabric of Cairene needle-work, the outside being waterproof and "wind-tight." To the right, as we sit, yawns the cooking-tent. In fine weather, the construction—I cannot dignify it by the name of "range"—stands outside. It is a series of holes with a slight sheet of iron built about them, and each glows with a charcoal fire. At this hour, the cook—Johannes,—Johanen ("Yohanen")—or, as we prefer to know him, plain John—presides over these craters, like a burly priest above sacrificial fires. Before I had learned wisdom by experience, I used to watch doubtfully processes so unlike any other methods of cookery with which I was acquainted as to waken apprehensions as to the outcome. That, without other utensils than two or three

JOHN AND IMBARAK.

pots, a couple of kettles, a frying-pan, a gridiron, a chopping-tray and knife, and a few spoons, he could get up a decent meal, might well tax the credulity of a Middle States housewife. That from the enchanted craters or retorts will be sent to our table, at six o'clock, a six-course dinner, as well-cooked and as daintily garnished and served as if furnished by Delmonico's chef, is a fact stated upon the authority of all who have partaken of these magical repasts. Excellent soup is the first course; two dishes of meat, always chickens or partridges, a "made dish" or *entree*, salad, pudding, or tart or custard or blanc mange, fruit, nuts and raisins and black coffee—is a bill of fare that represents our every-day family dinner. When a holiday intervenes, or distinguished guests are expected—ah, then, John of the many names buckles on his armor of proof and spreads a table in the wilderness that would have made David the Royal open his eyes in amazed delight.

"THERE IS OUR CAMP."

Next to the kitchen comes the dining-tent, furnished with table, and jointed or camp-chairs of Jamal's own design, having each four stout legs, perpendicular and trustworthy, and a good back of its own. Besides these, are camp-stools that may be used as foot-rests, and steamer-chairs or lounging-chairs for lazy and weary hours. The table furniture is that of an elegantly appointed home; the invariable brightness of the silver is a despair to one housewife who, after many years of patient effort, confesses herself unequal to the task of training hirelings to keep urn and teapot and spoons up to looking-glass lustre.

Imbarak dwarfs the tent every time he enters by his slim altitude of six-feet-three. He serves each course faithfully, omitting not one jot or tittle of the waiter's duty, and moves like a long-drawn-out shadow.

The number of sleeping-tents is regulated by the size and material of the party. That allotted to the only lady of the present company is in nothing more luxurious than that occupied by Alcides. It is spacious, and I can stand upright in any part of it. On one side is a dressing-table and behind it hangs a mirror. The cot-bed is most comfortable, supplied with feather pillows, clean linen, white counterpane and warm blankets. The floor, of earth beaten hard and smooth, has two coverings of carpet, the uppermost being Oriental rugs, soft and pretty. The polished pole in the centre is set about with hooks and rings on which wraps and extra clothing are hung.

I am the more explicit as to the details of our present abode because of letters

AFTERNOON TEA.

lately received from friends over the sea, sympathizing with me in the hardships of camp-life, and fearing that I may be permanently injured by the experience. One dear friend writes:

> "I cannot express my admiration of your courage in undertaking to dwell in tents for whole nights together. It is well enough for men, especially the young and healthy. For women, and those who are—well! to say the least, not as young as they once were—it is, to my way of thinking, imprudent and hazardous."

There are hundreds in America who would echo her sentiments, who, but for dread of the difficulty and discomfort attending Palestine camp-life, would gladly extend their journeyings to the oldest and most interesting of lands. For the benefit of such, I am describing our encampment just as I see it, this hour. In rainy weather, if we are caught abroad, we find our quarters snug and dry, if somewhat confined as to sitting room space. If the torrents are heavy, trenches are dug about the tent to lead the water off. The floors never become wet.

Dinner over—and we discuss it as leisurely as we like, time being no object in the evening—our lounging chairs are set—if the night be dry—without my tent door. To our left, the mules, eight or ten in number, and the horses, are pitched on the extreme verge of the encampment. Nearer to us, the muleteers, "the boy," (of thirty or thereabouts, whose name ought to be spelled as it is pronounced—but probably is not—"Serkeese,") and the grave-eyed, gentle-voiced sheik (pronounced "Shake,") sit upon the earth about the camp-fire, where they will sleep to-night. We wish vainly for an artist, or for a camera warranted to report by firelight, that might preserve for us the picture made by the unconscious group. The red light flames fitfully upon the bearded faces framed by the silken handkerchiefs (kafeeyahs) wound over their caps and under their chins, touching the gilt embroidery of the sheik's vest and the butt of his sword, bringing out the red scarf of one man, the yellow "kafeeyah" of another, and showing the intent eyes fastened upon the narrator.

SERKEESE.

We call up David to translate a few sentences of the story. He laughs in indulgent amusement at seeing that Serkeese is the speaker, and bends an attentive ear to the tale. It is, we learn presently, a story of a Sultan's son and a fairy who is enraged because he has broken her water-jar. "I cannot punish you because you are the son of my Sultan," she says, "but you shall fall in love with a beautiful woman who lives seven mountains away."

"It is a scene right out of the Arabian Nights!" cries Alcides, delightedly, and we fall anew to watching the countenances of the auditors.

Serkeese uses his hands freely as he goes on; sometimes he is very animated; the attention of the listeners never wanders from him; now and then a low exclamation evinces the feeling aroused by the simple—to us childish—recital of love and adventures, of dangers many and miraculous deliverances. Men fifty years old will hang upon the utterance of such until midnight; occasionally they hearken immovable all night long.

We have our own talk in the dimly-lighted background, while the bands of Orion glitter in the zenith and the nebulous glow of the Pleiades brightens in the darkening night, and the sacred mountains keep watch and ward upon the horizon. Often we ask David to join us and tell us stories of real life; of Bedouin prowess and Bedouin thievery; of Arab love and Arab murders; of Moslem manners and customs and superstitions. He has them all at his tongue's end, and if urged will give us, less freely, incidents of his own early adventures, when travel in the Holy Land was another name for peril, It was not possible, then, for foreigners—or natives, for that matter—to go from Dan to Beersheba with no other protection than a mounted sheik as a pledge of nomadic tribes that all is right, and, according to their tribal code, lawful. Armed men by the score attended the travelers, and attack, in certain regions. might be expected at any instant.

"What has made the change?" we ask him.

The loyal Syrian draws himself up proudly.

"The present Government, madame!"

We laugh, having anticipated the reply.

"Why, David, you will convince us, by and by, that your Sultan is a pretty good sort of a fellow."

"We never had a better ruler, madame. Look at the improvement in our roads, and how it is so much safer to go from place to place, and what he is doing for the poor. He is most merciful and tender-hearted, moreover. He has given from his own purse twenty thousand pounds toward rebuilding the mosque that was burned at Damascus.

"Then, too, there is Abraham Hakee Pasha, Governor of Jerusalem, who gives much attention to the condition of the city and the poor and needy. A very just and upright man in all his dealings. And El-Haj Saleem-Effendi-el-Husane, head of the municipality of Jerusalem, gives all facilities for the welfare and comfort of the inhabitants. A noble gentleman is he. His parents are of the oldest house in Jerusalem. I cannot but speak well of our rulers and governors, madame. They are doing their best for us. And madame must see that there is much to be done."

He is so much in earnest that we refrain from expressing our gratification that the sheik of the neighboring village has sent a guard to patrol our camp to-night.

Whenever it can be done, the encampment is within a few minutes' walk of a hamlet or town, whence a guard of two men is drawn to be guarantee to the government for the safety of the travelers. So secure does this precaution make us that we lay us down to sleep as tranquilly as if we were behind walls and bolts at home. If additional surety were required, we have the knowledge that David Jamal makes the rounds in person twice during the dark hours to assure his faithful soul that all is well with his charges. I have never slept more soundly and

READY FOR THE MARCH.

healthfully than in his tents. If we awake once in a while to hear the queer cry of jackals over their prey, or are startled by the weird laugh of a hyena and the bark of a wolf, we smile at the oddity of the situation and turn us again to our pillows with renewed consciousness of safety in our guarded nest.

Morning comes all too soon. Usually, the first knowledge we have of the return of the light is brought through the scratch that serves as a knock at the door of the tent, and the civil sentence in Arabic that signifies "hot water" in English. The smoking can for our toilet stands at the entrance.

No refinement of civilization that thought and pains can secure is lacking. Ours is the universal continental breakfast, also in vogue at the East—coffee and tea, bread and butter, eggs in some form, and marmalade, honey or jam. While we eat it, all the tents except that in which we sit are quietly "struck" and packed. When we come out, horses and men are ready, and the gravely-courteous morning salutation of our dragoman is coupled with the reverent,—"By God's will we will see" such-and-such a place "before the sun sets again."

The sheik swings into his saddle, and leads the way in the direction taken by the mule-train; Massoud, the wiry gray ridden by Alcides, curvets and kicks like a merry colt in falling into line; Serkeese rides briskly by, a grotesque figure upon his perch atop of the roll of luncheon booth and hampers; before we have gone a hundred yards, David, who has tarried to issue a parting order to the three men engaged in clearing up the debris of the camp, gallops to my side, his mettled charger, Dervish, brave with embroidered saddle-cloth and fringed housings, and we are fairly started on the march.

CHAPTER VII.

THE NATIVE GIRL.

AT my first interview with her, I, for awhile, mistook her for a boy. She sat flat upon the floor, her legs crossed under her, and her head uncovered. Her body was clothed in a pair of cotton trousers, and a blouse of the same, belted at the waist. Besides these, she had on not another garment. Her eyes were black, her face expressive; her gestures were animated, and she talked a great deal. She was, apparently, about ten years old, and well-grown for her age.

"A lively little fellow!" said I to my interpreter, after watching her for some amused minutes.

"She is a girl, and the terror of the neighborhood," was the rejoinder. "One who in your country would be a brilliant woman in time, maybe, with progressive ideas. Here—poor child!"

It was needless to finish the sentence. I knew enough already of the life of a girl in the East to be able to lay in the shadows of the scene.

"Can she read?" I asked.

"Not a word—and never will know her letters."

"What training has she in housework?"

"None. Her mother will tell you that she is lazy. She carried the baby—there is usually one every year—until a younger sister was big enough to shoulder the load. Then our little friend here 'struck' playing nursery-maid. She cannot sew a stitch, or knit, or cook, or wash, or iron. She runs wild all day long; she can out-swear her brothers—and there is nothing that should be hidden from a young girl that she has not known since she was four years old."

The speaker was gentle of speech and of judgment, but I was inclined to hope that she was severe upon the subject of the unflattering sketch. Subsequently I learned to know Lala well, and recognized the fidelity of the portraiture.

I have told how unwelcome our baby girl was when she first made herself heard in the one-roomed abode of her wedded parents. It is doubtful if Lala's father has ever given her an affectionate word. If not actively unkind to her, he submits to her presence in his household as an inevitable disgrace. It matters less than nothing to him that she grows up like a weed, or like a "tramp" dog or cat. If passably good-looking, she will be married soon and out of his way.

She stands within sight of me as I pen these lines. To be fleshy is considered an attraction in a woman by her countrymen, but Lala (Lai-la) is as yet slim. She is also straight and well-formed, in spite of the circumstance that the eight-months-old brother she carries upon her hip, encircled by her arm, is her charge when out of the cradle. Upon the back of her head is pinned a white veil—cotton and coarse. This she has been compelled to wear in the street from her eighth year. Seeing mere babies with veils on, which they coquettishly pulled

MRS. DAVID JAMAL'S "INDUSTRIAL CLASS" OF SYRIAN GIRLS.

over their faces at meeting men or strange boys, I for some time supposed that they were worn of choice, as little girls with us masquerade in grown people's clothes. I discovered later that every prudent Moslem mother insists upon thus early investing her daughter with the badge of womanhood. When we come to comprehend what womanhood is in this land, we feel that the badge should be crape or sackcloth.

"Do you love your little brother, Lala?" is the beginning of my dialogue with her.

She laughs—the childish, half-bashful, half-pleased laugh one often hears here from the mother of several children—women who never mature in intellect and whose physical maturity is that of summer fruit that has a worm at the core.

"I don't know."

"Is he heavy?"

"Oh, a little—sometimes," indifferently.

"How old is he?"

She looks at him, as if the thought were new, and laughs again.

"I don't know."

"What do you do with yourself when he is in the cradle?"

She nods backward at the door. "Go out into the street."

"To school?"

This is the funniest idea of all, and makes her shake her head many times.

"No! no!"

"Can you sew?"

She lifts one brown hand, shapely and slender, we observe, and seems to examine it with infantile curiosity.

"No!" is the result uttered slowly.

"How should she?" says the interpreter aside—that is, in English. "Her mother does not know how to use a needle. Poor as she is, she hires a sewing-woman even to do what family-mending is absolutely needed."

As I jot down the reply, Lala, having given the baby to his mother to be nursed, sinks upon the matting beside me, and fingers lightly, first my sleeve, then the skirt of my dress, which has a band of fur on the bottom. I cannot but smile into the child-face. I now perceive that it is comely, and would be intelligent had the brain back of it ever been awakened. I lay my hand upon the inquisitive, yet respectful fingers, as they pass to my stockings.

"Would you like to wear shoes and stockings, Lala?"

She laughs three or four times, and nods twice as often, then explains complacently that when she is a bride she will have a pair of stockings—perhaps several pairs—and sandals, besides other fine things—new gowns and bracelets—holding out slender wrists encircled by strings of beads, blue and white.

The word "bride" is talismanic. When a baby-girl falls down, or hurts herself in any way, the mother or neighbor checks her lamentations by calling her "little bride," as nurses in America "fie-fie" their charges with "Little ladies ought not to cry!" If the mother would cajole the unruly daughter she has never controlled by reasoning, she wheedles her with, "Come now, little bride, do so and so." Her narrow, pitifully starved life has but one ambition—to be married. She knows no other game than playing "bride," when her hair is dressed with flowers, herself covered with a veil, and a play-fellow takes the part of bridegroom. For this end she was created and has breath and being. I put up notebook and pencil and look at her, as she is—a type of tens of thousands in this so-called Holy Land—in the sight of women who have, all their lives (as have their mothers and great-great-grandmothers before them), enjoyed the gifts of Christianity to our sex without appreciating the depth and height, the length and breadth, the unspeakable value of what we owe to it—with pity for the ignorance I cannot enlighten, the degradation which nothing but the religion of the Christ who humbled himself to be born of a woman among the Bethlehem hills over yonder can relieve.

Hampered as I am by the absolute impossibility of putting into print that which might make the very paper on which I write blush for shame, I writhe, as in bondage, because our wisest and best American Christians are not familiar with so much as the general features of a woman's life here and elsewhere in Eastern countries. What I cannot write, and what you, my sisters, would not let your daughters read, is a part—and a large part—of the everyday life going on under my eyes. Our home-missionaries see much of what is popularly termed "the seamy side" of human living. The missionary in the foreign field might cry out in bitter disgust of spirit that he sees nothing but seams, and darns, and patches in garments stiff with filth.

Mrs. David Jamal, the sensible Christian wife of our dragoman and guide, was educated in a Beirût school for girls, and went to Jerusalem as a native helper in a school there. Since her marriage, as before, the condition of her fellow-countrywomen has been a sore weight upon her mind, and, although the mother of a large family, and a diligent housewife, she, some years ago, collected in her own home a class of native girls and began to give them instruction in plain sewing, knitting, crocheting, embroidery, and in all departments of housework. Twice a week she has presided over what, in her humility, she never thought of calling "an industrial school," the attendance varying from thirteen to eighteen. She can accommodate no more than eighteen, and has always applications on hand from ten or twelve who wish to join the class, and cannot for want of room.

Mrs. Jamal gives her time and house, and, up to this time, the expenses of materials for the work and the luncheon cooked twice a week by the girls under her supervision have been met by the sale of articles manufactured by the class, and an allowance of twenty dollars a month made by a benevolent man, a sojourner for some years in Jerusalem. He has now left Syria, and other circumstances make it impracticable for him to continue the appropriation to this branch of mission work. Unless help comes from other sources, this most sensible and practical enterprise of a Syrian woman for the benefit of her sex will have to be abandoned.

I wish it were possible for some of the societies represented by my readers to take up the cause of this industrial class. Twelve "circles" contributing each

COASTS OF TYRE AND SIDON, BETWEEN JAFFA AND BEIRUT.

twenty dollars a year would enable Mrs. Jamal to carry it on and to receive a large number of pupils. I may add that, steadily and tactfully, she infuses religious teaching with instruction in handiwork and housewifery.

Upon a grander scale, and aiming steadily to accomplish the same end—the elevation of Syrian women from fellowship with the beasts that perish to the plane of rational, immortal beings—are such schools as Miss Everett's in Beirût; that at Sidon, in which Mrs. Dale, the daughter of Dr. Bliss, of the Syrian Protestant

"WOMEN WHO ARE NEVER MATURE IN INTELLECT."

College, is an efficient laborer; and others at Nazareth, Haifa and Bethlehem under the care of English societies.

One and all find a serious impediment to their work in the infamous system that gives children of ten, eleven and twelve years of age to men whose conceptions of the nuptial relation are of the lowest order. Still, there are parents who, through ambition or natural affection, desire that their girls should be educated like European and American women, and allow them to remain under the care of Christian instructors up to sixteen or seventeen before transferring them to their father's or husband's houses. For every such exception, we thank God and take courage.

My pen was in place to affix the signature to the above when the sound of chanting in the street drew me to the balcony. Down the middle of the muddy thoroughfare walked a short procession of men in striped "abiehs" (native cloaks made of coarse camel's hair), chanting intermittently and listlessly, followed by a troop of ragged boys. The foremost man bore in his arms the uncoffined body of a baby. It was wrapped in a white cotton shroud; the waxen face and closed eyes were pretty and peaceful. The sight of a dead infant always wrings my heart to breaking, and the old, sadly-familiar ache forced from me an ejaculation of pain. "Oh, the dear little baby! and the poor, poor mother!"

"It is a girl!" said a voice in my ear. "Thank God it is dead!" An American woman, whose nature is all sweetness, and her guest, an English teacher in the Bethlehem school, were in the adjoining balcony.

"It is taken from the evil to come," pursued the speaker, answering my look of inquiry. "If you knew, as we do, the almost certain doom from which that little creature has been saved by death, you would say, 'Amen!'"

CHAPTER VIII.

FUDDA'S BETROTHAL.

IT is something to get fast hold of a woman in this country who does not take fright at the sight of note-book and fountain-pen. It is something more to secure one who knows of herself just what I wish to find out and to report. Best of all it is to discover that my brave woman, my well-informed woman and the woman who can speak Arabic fluently and English graphically, are one and the same, and cheerfully at my service.

I have seated her in my room and given her due notice that I mean to make use of her to the utmost. Circumstances—not the least formidable of which is the accident of my foreign birth and breeding—have kept me, to some extent, in the outer court when certain ceremonials are performed in the innermost of the home. I may see marriage processions, but as a visitor, not a participant in the festivities and solemnities of the occasion, and when I have sought to "interview" prospective brides and brides' parents, I have been conscious that a meagre outline of history is all that I can obtain.

But here is one who was born and brought up in Syria, who has used her eyes, ears, tongue and brain, who has a fine sense of humor and the reportorial faculty of seeing everything, and telling what she has seen. By virtue of some or all of these advantages, she has assisted at a dozen or more Moslem weddings, and has one fresh in her mind. I shall let her tell the story as nearly as possible in her own way. The occasional lapse into native idioms adds flavor to the narrative. I only regret my inability to enliven it for my readers as she brightens it for me by gesture and facial play. The nervous action of her strong, supple hands, the flash of eyes and teeth, as she touches the comic features of the story, would make tame language piquante.

She begins apologetically:

"You will not mind that I say how strange" (rolling the r's richly) "some of the European customs of young men and young women appear to us. I have seen young people who are betrothed walk in the streets, arm-in-arm, and when they do separate, say, for the gentleman to go into camping in the interior, and the lady to remain here with her mother or friends, he will take her in his arms and *keess* her in sight of others! And these are not common or vulgar or not-educated people. Oh, no! but ladies and gentlemen of the best order in their own land—genteel and religious. The ways of these strangers are wonderful, and may belong to the customs and climate of their countries, but they would not be suffered here—not for one hour. With us, if a girl smile so much as once at the man she is to marry, he will not have her—he will cast her off. He will say —and he is wise—'If she be light in her behavior to me, she will be light to any other man. She is not to be trusted.'"

"Yet they must become acquainted with one another, before they are married, and it is natural for a light-hearted girl to smile while talking to a lover."

"She never talks to him!" hands and eyes combining to enforce the declaration. "Usually, she has never seen him unless she catch a glimpse of him through the window, when there is not danger that he will see her peep at him, and despise her for being bold and curious.

"A CHILD, AND NOTHING MORE."

"You see it is this way: A man who has work, or money from his father or in some way, says to an old woman—perhaps it is his aunt, perhaps it is his cousin, or perhaps it is an acquaintance,—'It is time I get married. Do you know any girl that would suit me?'

"She is always on the outlook, you see, for it is many presents some get in this way, and she has ready an answer:

"'Oh, yes! there is (we will say) 'Fudda, whose father lives near by me. She is a pretty girl, and young and healthy, and of a good family.'

"Then she tell him who is her father and her mother, and how respectable they are, and how young and good-looking is the girl. Perhaps she is but ten years old, or she may be fourteen."

"I know," I interrupt, "your girls mature much earlier than ours. A girl of twelve with us is a child, and as for one of ten, she is hardly more than a baby."

My companion is grave enough now; her eyes are dark and moist; she sinks her voice almost to a whisper.

"A girl of ten is a child with us. A child and nothing more.

"Yet I know one who is married to a man seventy years old! Her father could not give her bread, he was so poor, and this old man gave a good price for her—fifty napoleons, for she is a pretty child."

"A good price! Do fathers sell their children?"

"You shall hear. We do not call it that, but when you have listened, you may say it is what you please. Well! the young man is willing, after what he is told, to marry the girl, and the old woman goes off to see Fudda's mother, just as if she had nothing of importance to say, but was passing by and thought she would make a visit, and presently, when they have had coffee together and smoked a few cigarettes, she says, if Fudda is in the room, 'Will you send your daughter out? I wish to speak with you.' Then she will bring forward the name of the young man, and make many praises of him, and tell how he would wish to marry her daughter. The mother knows, and the old woman who goes between knows, that all depends upon what the father shall say, but they like to talk it over for a long time, maybe two hours, and back the old woman returns to the young man, and says, 'I have made straight the path for you so far as I can. Now it is for you to act.'"

"I see!" (and I think that I do.) "He must speak to the father and ask permission to address the daughter. That is the custom in some European countries, as well."

It is plain that I am not upon the right line. Her gesture is deprecatory.

"That would not be correct. It would be too much haste—and not at all respectful. The young man goes to his uncle or his brother—if he be an older brother and very sober and—and—deegnified, and says—'I want that girl. Will you talk to her father and see what arrangements you can make?' So the friends—for it is quite usual that he sends two—go to the father and tell what a respectable young man this is, and how respectable and of good family are his parents, and that he *weesh* to marry his daughter. Perhaps, it may be, that the father is not willing, for some reason, to let the marriage take place. Then he laughs and puts to scorn all that has been done. He say—'This is some silly gossip of the women. I will give my wife a sound scold for her foolishness. You may go tell your nephew—or your brother, as it may be—that I have no daughter to be married.' And there is the end of it all."

"But if the father has no objection?"

"Ah! then it goes on. The father says, 'How much will your friend give for this, my daughter?' Then they bargain and bargain. If she be pretty—very pretty—the father may ask as much as a hundred and fifty napoleons. But it is

(90) FUDDA AT THE WELL.

likely that he will not get more than eighty, or maybe seventy-five. If she be not handsome, maybe he will get but forty. This, when the father is not rich, or when he is greedy and wants the money for himself. You see it is this way: He say, 'I have brought up my girl and had the expense and trouble of her all these years, and she has not earned for me one piastre. Her husband must pay me something for all that.' But the better class of Moslems, and especially of our own people—the Christians"--(this includes Roman Catholics, Greeks, and Protestants), "will prefer to have the price paid in jewelry and clothes that will belong to the bride. Then her father will have to give her nothing from his own house. So—the bargain is made, and the bridegroom and his friends, and the uncle or brother of the bride—and if she have no uncle or brother, perhaps her father, will go to a judge—an advocate—do you call him?"

"A lawyer—do you mean?"

"That is what I would say. They go to a lawyer, and he writes down on a paper how that this man—mentioning his name—desires to have in marriage Fudda, the daughter of such and such a man, and promises to give on such a day, first a ring (the first present must always be a ring), and on another day a silk dress, and again a gold chain, maybe, or a bracelet, and next another dress, and to this he signs his name, and the witnesses write theirs, and the lawyer signs with the seal, and the paper goes to Fudda's father. This is the 'cere-mo-nee.'"

"I beg your pardon! What did you say?"

"The 'cere-mo-*nee*.' This is what makes her betrothed to this man. There is no other. No church service. No questions, no answers. No promises. No vows. As soon as what the bridegroom has promised is paid, he can claim his wife. He can go to her father and say, 'I have here your receipts for so much money—or so many dresses and jewelry. Now I will have my wife.' She is his by the law, and if her father did not give her up, the law would take her away from him.

"Once in a great, great while, it happens that the man has promised what he cannot pay, or he is careless, or he is not honest, or he does not care so much for the girl as he did at first. Something like this makes him slow in keeping his engagement. One, two, three, maybe four years are gone, and he has not paid all, and Fudda is living in her father's house, his wife and yet not his wife. By-and-by, the father loses his patience, and he sends word to the man—'This must come to an end. My daughter cannot be married to any other man while she is betrothed to you, and you cannot have her until you have filled your part of the contract. If you can pay what you have promised, do it at once. If you cannot, or do not mean to do it—if you are so disho-no-ra-ble as to wish to put this deesgrace upon me and my daughter—then divorce her, so that I can give her to some better man.' If he cannot or will not keep his engagement, he must divorce her, just as if she were his wife."

"Suppose the poor girl has become attached to him in this time?"

"She could not!" This is said so positively that I am somewhat abashed at having offered the suggestion. "She has never seen him, unless she may have met him, by chance, in the street. Then it is her duty as a modest woman to turn her head aside"—suiting the action to the sentence—"and hurry by. Of course, he has not once seen her face, for she never stirs out of doors without her veil, and should he come to her father's door on an errand, she must hide in a cupboard, or run behind a curtain, or con-ceal herself.

"Suppose, though, the young man has kept his word, and all the price is paid, and the lawyer writes his receipt that so many jewelry and dresses and so much money has passed through his hands to Fudda's father. Sometimes this is done in a few weeks, sometimes in a few days. If the time is short, the girl does not know that she is to be married, or that any man has asked for her, until, maybe, two or three days before the wedding. Then, her father will say to her—'On such a day—maybe next Wednesday or Monday—you are to be given in marriage to such a man.'"

"Whom she has never seen!" I shudder forth.

"It may be that she has never heard his name until that minute."

"What if she refuses to marry him?"

"Ah, then, would not her father give her a good slapping? He would ask her, 'Who knows best what is good for you—you, a silly, ignorant child, or your father and your mother? You disobedient, ungrateful girl! And if you would not have your father's curse, and be turned in deesgrace from your father's house, do as you are bid.'

"And, indeed, madame," interjects my informant, judicially, "he is in the right of it. What can a girl who has been kept close in-doors when there is any danger of meeting a man, and seen nothing of the world, understand of such things?

"Are such marriages generally happy?"

It is not a query to be dismissed lightly, and she ponders a minute.

"Say there are one hundred marriages made in this way. I should say that ten out of the hundred are not happy. And"—slowly, and again dropping her voice—"I have, myself, known of two girls who threw themselves into a well to escape marrying men chosen by their parents."

David Jamal has allowed me to read an interesting paper prepared by himself, upon "Marriage Customs in Jerusalem and the adjacent villages." It contains, among other valuable and curious things, a literal translation of an Arabic song sung by the "future bridegroom and his friends and relations" at the betrothal-feast in the house of the bride's father.

At my request, Alcides has written a metrical version, that yet faithfully renders the spirit and almost the language of the original:

O father of the bride! throw wide each door
  To welcome us, the long-robed bridal train.
May Allah, in his mercy, evermore
  Protect thy home from every grief and pain.
Thy daughter's fame has spread to distant lands,
  Men call her gracious, fair and swift of thought,
Her father, prince of gallant desert-bands,
  Whose wealth Aleppo's castle might have bought.

O father of the bride! be generous,
  Ask not a heavy dowry for the bride;
Thy gold will perish if 'tis gathered thus
  But we, thy friends, remain forever tried.
Then welcome us, thy friends, the wedding guests,
  Prepare a feast for us beneath thy dome;
And then, when thou hast granted our behests,
  Mount thy swift steed—escort us from thy home.

CHAPTER IX.

FUDDA'S WEDDING.

AND I must tell you, at the first, madame, that her father is a rich man and much respected, and the man she was to marry is also rich. You must know that the better the class of people, the more strictly they follow all the rules I have spoken of, regarding betrothing and marrying. So I was very anxious to go to Fudda's wedding, for I had heard much of the beautiful presents she had received from her bridegroom, and the great company that was to be, and the feasting, and all that.

"When the mother of the bridegroom heard what I wished, she was very kind. She said to me, 'You shall come with me, and see all at my side.'

"She was as good as her word. When the night of the wedding came I was beside her in the procession that left the house of the bridegroom's father. Such a procession! The women in white 'izzars,' and veiled, some with crowns of flowers on their heads, and the men in their best clothes; the friends of the bridegroom and the near relations of the bride in wedding-garments that were presents from him. But oh! it is an expensive matter to be married here!

"I have heard sometimes, when the uncles and cousins or brothers of the bride are not well-satisfied with such gifts as he has sent to them, they will stop his way to the bride's house, and cry, 'You shall not have her until you give me a coat,' or maybe trousers, or something else.

"No such thing happened the night I am telling you of. We marched—the women all together, and the men together, and many carrying torches, and a band of music playing merry tunes at the head of the procession, until we came to the house of the bride's father. There the men stopped, at the gate and in the courtyard, and we women went on up to the chambers. One of them was locked, and inside of it was the bride, and no one had a right to open that door until such time as the bridegroom's mother should come. Everybody made way for her, and you may be sure that I kept close beside her. She unlocked the door and threw it open, and oh, dear! what a rush there was!

"The first thing I saw was Fudda, seated up high in a chair, and the chair set upon a sofa at the back of the room. She was covered all over with a white izzar, like mine,"—pointing to the sheet-like drapery she had laid aside at my invitation—"and a *mendeel* like this, over her face,"—holding up the square of flowered tissue edged with hand-made silk lace, always worn by Syrian women in the street—"and over all was a long veil of pink muslin, reaching down to her feet. There she sat, straight and never moving, her hands in her lap, and her eyes closed under the two veils. You know that the bride must go to her husband in pink?"

"I thought white the bridal garment all over the world."

"Not with us. Pink she must wear, or the marriage will not be fortunate. On the floor, directly in front of Fudda, was a stand, and upon the stand a great brass tray, as big as this,"—extending her arms to indicate a wide circle—"with ever and ever so many candles lighted upon it. I thought there must be one hundred. All the married women rushed for a candle, and there was screaming and laughing and scrambling such as I had never heard before. I was afraid I should be knocked down and trampled upon.

"And what do you think it was for? Every married woman who got a candle would see her own children married in her lifetime and in her husband's lifetime as well. Of course, that is all superstition, but the Mohammedans are very, very superstitious."

"What was the bride doing all this time?"

Teeth and eyes make arch play over the comely face. "Sitting still and her eyes shut! It is not good fortune to open them. Presently, when the candles were all taken, in came her father and her brother, and helped her down, and put on her two swords, one belt crossing her breast—over the veil so, the other so!" designating the right and left sides. "Next, they led her down stairs and out of doors, she sobbing behind her veils. You could see her tremble and shake all over with grief at leaving her home forever. At the door, up stepped four men carrying —what do you name a cover fastened up high on four sticks?"

"A canopy."

"That is it! And under this canopy she walked in the procession to the house of the bridegroom's father. She walked, oh, so slow, about one inch to each step—so!"

She rises and folding her hands and closing her eyes, marches, stiffly and laboriously, deliberately across the floor. We both laugh heartily, as she resumes her natural gait and manner.

"She must have looked like a jointed doll," I say.

"That was what she did look like- and the joints were not easy. I thought we would never, never get to the end of our walk, and the band played all the time. At last, we were there, and again the men were to wait outside, while the women went in. The bride was taken with us. She did not seem to go in of herself. They sat her down flat upon a rug on the floor, and took off the swords, and the pink veil, and the short veil, and the white 'izzar,' and we could see that she had on only her little girl's dress of clean print, made like a child's. She did not look to be more than eleven at most, and like a baby that is scared, her eyes shut, and her face wet with tears, and biting her lips to keep from crying out aloud. She kept very still, however, while they stripped off her print gown and stockings (her sandals were left at the door with all the rest), and put on her finery. Silk stockings and silk trousers and a fine—very fine—pink silk frock. And the women crowded and talked, and were busy all around her. One would put on her stockings, and one braid her hair into two long braids that hung down behind, and were tied with pink ribbons. Another put her dress over her head, and one fastened it, and one stuck flowers in her hair, and this one pulled a pair of long gloves on her hands. All would take a share in decking her out in her jewelry. I never saw so many bracelets and chains and brooches upon one person before—diamonds and rubies and emeralds and turquoises. When her gloves were on, the rings were slipped upon her fingers over them. I could not help laughing to myself to see her—always keeping her eyes shut—stick out her fingers, so, to have the rings put on, just like your doll with joints. When she was all dressed, the women pulled her upon her feet, and a pink veil

"IT WAS PLAIN TO BE SEEN THAT HE WAS PLEASED."

was fastened to her head and came to the floor in front. Then they brought again the two swords. Now they were crossed—in this way"—making a St. Andrew's cross with her forefinger—"and tied with ribbons. Two women held them over her head, an old woman who is a sort of doctor and wise woman, you know, and another as old and much looked up to. These swords signified her father and her brother. You see, should her husband not be kind to her, he must recollect that she had somebodies to protect her.

"Now the door was opened and the bridegroom was waiting on the outside, and he came up to her where she stood between the two women, with the swords crossed over her head, and the pink veil down to her feet, and the jewelry shining through it, and on her arms and hands. Then he lifted the veil and saw her face-to-face for the first time in all their lives. It was plain to be seen that he was pleased, and he took her by the hand and led her to his own room.

"It is not what you call the etiquette for him to stay there more than a few minutes. He came out soon and took his part in the feasting and dancing and making merry that lasted all night long. The next day there was more feasting in the house of the bridegroom's father, and the third day yet more, all at the expense of the bridegroom. I have known weddings that cost so dear that the bridegroom had to sell some of the jewelry that he had given his wife to pay his debts. The bride is not seen at any of the feasts. From the minute she is a wife, she must not speak to another man except her father and brothers. Not so much as her husband's brother, unless he permits it. And he does not often allow his brother to see her, for fear she should like him better than she likes her husband. He would divorce her in one minute if he found her talking to any man, even through a window very high from the ground."

"Could he divorce her legally upon such a pretext!"

"He can divorce her for anything he likes. If she has not his supper ready in time; if she is not respectful to him; if she does not come quick, on the minute, when he calls her; if he would like to marry a younger woman, he can say the three sentences that mean, 'Go! I am divorced from you!' and she is no more his wife.

"When he speaks these three sentences, she will drop whatever she is doing—sewing, washing, cooking, nursing her baby—and draw her 'izzar' over her face, and go straight out of his house. She must not look at or speak to him any more than to a strange man. Perhaps he has been angry and said more than he meant, and is sorry in a minute, and tries to speak to her, or would follow her. 'No! she says, keeping her face hidden and putting out her hand to warn him back—in this way:"—a dramatic gesture—"'You are not my husband. I do not know you!'"

"Where can the poor creature go?"

"To her father, if she has one. If he is dead, to her brother, or if she has no brother, to her uncle. He says to the man who was her husband: 'Give my daughter, or my sister, or my niece, the clothes and the jewelry that belong to her. My house is her home until she can get another husband.'"

7

"And her children? What becomes of them?"

"If she has a nursing-baby it goes with her, and the father must support it until it can do without her. Then it belongs to him, like the other children."

"Cannot the mother visit them, or have them with her at times?"

"No. They are not hers any longer. They must go with the father."

"Does he often marry again?"

"Always! always! A man must marry, you know."

"I hope the new wife is kind to the children," I say, sighingly.

Then the narrator smiles roguishly, the look that makes her face attractive. Such a look, I may remark, as I have never yet seen upon the visage of a Moslem woman. These carry in every lineament the stamp of inferiority to the other sex: are mere mindless machines to minister to the pleasure and comfort of their masters.

"It happens, once in a while, that she is so cross to them, and keeps her husband's house so badly, that he sees plainly what a great mistake he made in divorcing their mother. Ah! but what would he not give if he had never said those three sentences? So he says to himself, 'I have made a poor bargain this time. I will have my first wife back again.' Now, what does he do but send word that she must be married to another man, and then be divorced, and marry her first husband again!

"Yes, it is true that I tell you. He cannot take back the woman he divorced, you see, but he can marry the woman another man has put away. Of course, it would not be a sensible man who would marry somebody he knew would be taken from him next day. So the woman's friends look about for a man who was born silly—what do you name him?"

"You cannot mean an idiot!"

"Yes! yes! that is it—or a crazy man. They pay him to take this woman as his wife, and to-morrow to divorce her, that her first husband may have his wish. It seems horrible to you, and we native Christians cry out upon it as a sin and a shame, but is thought nothing of by Mohammedans. It is quite usual for a man to marry one, two, three, four wives, and divorce them all. I know a man who had six, one after the other. He divorced four, one died, and the last one is now his widow."

"And the divorced wives were not disgraced in the eyes of acquaintances?"

"But, surely not! If they are young and pretty, or very good housekeepers, or if they have some money, it is as easy for them as for girls to marry."

BETHLEHEM BRIDE.

CHAPTER X.

AMONG THE LEPERS.

A LEPER IN THE SUN.

IT is, I think, incorrigible Mark Twain, who says that the name of the "street that is called Straight" is the only bit of irony in the Scriptures. The flippant fling comes into the mind of many pilgrims to the "Eye of the East." It arises to our lips as we follow it to the Eastern gate of the city, and see it merged into a wider and less devious road.

This is bounded on one side by the dry moat of the wall. We ought, after two months' seasoning in regions associated in the western imaginations with attar of rose and spicy breezes, and Araby the blest—to be coolly indifferent to the effluvia arising from the wasting carcasses of two camels which have been thrown out of the highway into the ditch. A pack of lean scavenger-curs, yellow and black, are snarling over the unsavory spoils. Skeletons, denuded and dessicated, lie further on, and small bones make an unattractive miscellany between the larger and more pronounced.

"It is an Eastern poet, I believe, who says that Damascus is as fragrant as Paradise?" observes Alcides, with his nose well in the air.

There is no reply to the gibe, for our guide, Abraham Ayoub, a Damascene, selected by David Jamal, "halts" us at a miserable gateway in a more miserable mud wall on the left of the road. If there were ever gates, they have fallen from the hinges long ago. This is the entrance to a filthy court-yard where the sun falls with kindly warmth. Four human figures are huddled in the sunniest corner, mere bundles of rags to the casual glance. Two look up as we pass, and our shadows fall upon them. With our minds full of Bible stories of "lepers white as snow," we should not recognize these caricatures of humanity as victims of the awful plague, had not David's graphic descriptions prepared us for the revelation.

Alcides has recorded in his note-book as our dragoman's diagnosis of the loathsome thing—"a combination of itch, chilblains and gangrene." No more apt language comes to my pen, after having seen scores of the wretched objects. Beyond the outer court, we pass between two huts into a second enclosure, about thirty feet square, surrounded by squalid hovels, all facing upon the open area.

They are of mud, they are without windows; they have one door apiece, and they are unfurnished, save for the ragged mats which serve as beds. Scraps of moldy bread, bits of bone and rotten vegetables strew the floor—the remnants of meals devoured here as the dogs without the walls are gnawing theirs. Of the odors flowing through the doors into the sun-filled outer world, I cannot speak. On this fine day everybody is outside except a few women dimly visible in the obscure interiors, and at these we abstain from looking as we pass.

A strange voice, both hoarse and shrill, hails us while we stand in the middle of the courtyard. It proceeds from the porch of a small frame house directly opposite the opening by which we had entered. The house is, we note, larger than the mud huts, and has two windows besides the door. A woman, or girl, for she must be under twenty, and might be good-looking but for her viragoish expression and a certain hardness,—I might term it a callosity of skin and features that betrays the blight which has fallen upon her—stands upon the upper step wringing out a cloth she has taken from a great metal pan. We have interrupted her day's washing, and for other reasons she is ill-pleased. Her husky scream is a question, for our guide answers apologetically what he translates to us in a hasty undertone:

HOUSE OF CHIEF LEPER IN NAAMAN'S HOUSE OF LEPERS.

"This gentleman"—designating Alcides—"is a great 'hakim' (physician) from over the sea, who seeks to help you if he can. We intend no harm."

She has thrown the cloth back into the tub with a despairing gesture, and rushed into the house while he is still speaking, appearing a minute later at the window:

"Go away!" she cries, her voice breaking oddly upon the high notes as a consumptive's fails when he is about to cough. "Go away! Allah only can help us, and he will not!"

"Her father is the chief man of the lepers," says the guide to us. "He was once a rich—one very rich—man. Now he is here—and old—and cannot leave his bed. His housekeeper, also old, and this, his only child, live here with him in the Naaman House."

For this loathly place (to borrow a word from a former generation), is nigh unto the grounds where are shown a few columns surrounded by a mud-and-stone wall, said to be the ruins of the house of the famous Syrian captain. The asylum is maintained by Moslem charity. Wealthy Mohammedans die and leave legacies for the support of the colony, in the hope that the act may bring repose to their own disembodied souls. Kind-hearted people in the neighborhood give them provisions, leaving them inside the outer gate and hastening away to avoid contamination. In spring-time a few of the stronger men are hired to do light work in the fields, and in harvest to glean after the reapers, gather beans, etc. A less humane measure is the permission granted to the least afflicted of the colony to bake in the public ovens dough kneaded by their diseased hands.

We had been led to believe the Naaman House a hospital for what are surely the most grievously-smitten of all God's human creatures, and are surprised to learn that this retreat is merely a hole in which they can burrow and hide and be fed apart from the rest of the world. Last year there were forty-two lepers here; now, we are told, they number but twenty-eight.

"What has become of the rest?"

The guide lifts his shoulders to his ears. "What must come to all of them—death. Each is a dying man even now. See!"

Pity unutterable gets the better of disgust, as we survey the scene. The lepers lie grouped and sit singly in the sunshine, most of them against the wall, seemingly regardless of our presence. Curiosity is dead within them. They do not even beg for "bakshish," as is the habit of their kind in the streets of the city and on the wayside. Some crouch in corners, their poor, marred faces supported by hands from which the fingers are literally dropping; others lie flat upon their backs, soaking in the warmth, eyes closed and jaws falling as if dead. Every few minutes one begins to cough spasmodically, and the example proving infectious, another and another will join in, until a horrid chorus that threatens to shake the loose frames apart fills the place. Two near us crawl away, still coughing in a weak, wheezy way, clots of blood dripping from their lips as they go. None of them walk, but all shuffle along on their haunches like beasts who have been hamstrung, or creep upon hands and knees. In some cases, the hands and feet are muffled in dirty rags. Turks are here in garments that bear tokens of old-time finery in shreds of tarnished gold lace and frayed embroidery, and Bedouins that are ghastly travesties of the tall, bronzed horsemen we see in the desert and among the mountains. They still wear the costume of their race, the suggestion thus conveyed of the free, active life forever left behind them accentuating their present misery.

One of these loiters nearer to us than the rest, his eyes and manner expressive of some degree of interest in us and our movements. The girl at the window swears furiously at him, as he obeys the beckon of our guide's hand and approaches.

RUINS OF NAAMAN'S HOUSE.

"Dare not to go near them!" she vociferates. "They are dogs and the offspring of dogs, as their fathers were before them. They come, not to give alms, but to make a mock of us and to take pictures of us!"

The last half-sentence is drawn forth by the sight of the kodak swung from Alcides' shoulder.

A magic word of two syllables—the first I verily believe that Arab and Syrian babies learn to lisp—has quickened the Bedouin's step, and he stands before us, well-made and heavily bearded, with comparatively few traces in face and figure of the fatal malady. He is about forty years old, and when he speaks his voice is faint and dissonant. His lungs must be well-nigh gone. He is quite willing to converse—thanks to the prospective "bakshish."

"How long have you been ill?" we asked by the mouth of the interpreter.

"Since I was fourteen years old."

"At what age do the symptoms first show themselves?"

"In men, about fifteen; in women, twelve to thirteen."

"Are you contented—and—" bringing out the last word with an effort—"comfortable here?"

"I am not unhappy. I have no work to do. I have my lodgings. I have my food. What more do I want?"

"But if you have no work to do, how do you pass the day?"

"I sit in the sun. I do nothing. Why should I work? Man works to get clothes; I have clothes. And for bread and bed, and a cover by night when it rains. These are mine."

"Have you amusements here?" trying not to glance around the mean enclosure and at the repulsive occupants.

He stares vacantly. The idea is too new and strange to work itself into his brain. From the groups distributed about the court odd murmurs are arising; mutterings in weak, raucous tones, like the growling and whining of beasts disturbed in their slumbers in the sunshine. The woman leans over the window-sill and devotes us to the vengeance of heaven and the abode of the fiend. The whole affair is an abhorrent nightmare, from which we are thankful to escape. Our Bedouin casts furtive looks over his shoulder at the malcontents, and will answer no more questions, however mildly put. The liberal bakshish dropped from a prudent height into his maimed hand does not pacify the daughter of the bedridden chief of the queer colony. As long as we can hear her, she is reckoning up the misdeeds of our ancestry and imprecating curses upon our heads.

Dante never pictured a fouler deep than this so-called House of Mercy, where beings of like form and nature with ourselves are decaying piecemeal, while the breath of life still animates their dismembered bodies. As I have intimated, this form of leprosy must not be confounded with that which overtook Moses as a proof of God's ability to work whatever wonder He wills; or Miriam and Gehazi and Uzziah, as punishment for sin, or that which drove haughty Naaman to bathe in little Jordan, at the behest of a prophet who would not even grant him a personal interview. This is hereditary in forty-nine cases out of fifty, and the recognized result of the sins of the forefathers. David Jamal and his Damascus friend agree that the malady is unknown at this day among the Jews, testimony which is corroborated by other authorities to which I have referred the question.

"It is clearly, then, not caused by want of neatness in this generation," is the grave conclusion of our dragoman.

Intermarriage among lepers are constantly occurring, and sometimes, although, David says, not frequently, children are born to the horrible heritage of disease and disgrace.

Instances are on record of attempts to avert the certain doom by removing the child in early infancy to the judicious guardianship of others than the parents.

"THEY OBTRUDE THEMSELVES UPON OUR NOTICE."

In spite of sanitary and medical precautions, the unmistakable signs make their appearance between the ages of thirteen and fifteen. The incipient stages are intense itching of the hands, extending gradually to other parts until the whole frame is a body of living death, a mass of putrefying sores and decomposing bones.

The Jerusalem lepers who are at large are, if possible, in a sadder case than those sheltered in the Naaman House at Damascus. They have, it is true, a village of their own below Siloam (whose "shady rill" and growing lily are sadly dismissed to the realm of hymnal romance), a wretched cluster of huts to which they resort at nightfall, but the streets are their home, and begging is the only means by which they earn a livelihood. They haunt places frequented by tourists and pilgrims, following Russian and Greek devotees in their periodical visits to the Virgin's Tomb, Gethsemane and the Mount of Olives. There may be in the Holy City about sixty of this class of professional mendicants.

Singularly enough, lepers are found in but four towns in Syria: Damascus, Nablous, Ramleh and Jerusalem. These are to them as cities of refuge when attacked in the neighboring villages. No longer with the warning cry of "unclean! unclean!" they obtrude themselves upon our notice whenever we walk or drive through the thoroughfares in Damascus and Jerusalem; stumps of hands or fingers, from which one or several joints are missing, are thrust into our faces, the horrid voice, unhuman in quality and inflection, calls for "bakshish," and the disfigured faces grin or weep, as suits the owner's mood and taste.

The lights in this dark picture are found in the Leper Home at Jerusalem, founded in 1867 by the Baroness Keffenbrinck-Ascheradin, and by her transferred in 1880 to the Moravians. The building stands upon a hill pleasantly visible from the Bethany road. The management is excellent, and all the appointments are those of a well-kept hospital. Divine service is held twice a week, and those who are too ill to attend it are visited separately in their rooms. The "house father" and his wife and the chaplain address themselves to their terrible task in the spirit expressed by the last-named in a recent report:

"Remembering always that the loathsome and terrible disease of the patient is one which seems ever to have thrilled our Saviour's breast with a keen and instantaneous compassion. . . .

"Gospel addresses, prayers offered at the bedside of many a stricken sufferer and private intercourse with Moslem and Christian patients have led many to acknowledge the benefits of Christianity and to endeavor to live according to the high principles of the Gospel."

Physicians and nurses acknowledge that the malady is absolutely incurable. They are likewise unanimous in the opinion that, however long the seeds may lie latent in the individual or family, leprosy is not to be eliminated from the blood. All that they can do—and this is much—is, by cleanliness, alleviative medicines and external applications, to lessen present suffering, and smooth the pathway to the grave.

Wild rumors of extraordinary cures effected by a German physician who was here for a while, and whose work included some skilful operations, and of the miraculous properties of a certain oil, have excited false hopes in many who eagerly apply for admission. Those in charge of the Home are firmly candid in dissipating these fancies, declaring that no mortal skill can do anything more than to alleviate suffering and to delay the horrible end.

CHAPTER XI.

"THE PEARL OF THE EAST."

OUTSIDE the walls of Damascus—which, if you like, you may consider the setting of the "Pearl"—we first encounter a couple whom we are to know so much better by-and-by that I may be excused for introducing them somewhat at length.

Luther Zwinglius Sharpe, D. D., is the ex-pastor of an Eastern church in America, and ex-president of a Western college, and these are but a few of his titles to the respectful notice of the world in general and his country people in particular. A Congregationalist by birth, education and choice, he yet committed the ecclesiastical solecism, five years ago, of marrying a High Church Episcopalian of the most advanced type. She was rich then, and has lately become much richer. Shortly after the later accession to her fortune, the Rev. L. Zwinglius Sharpe (thus he registers his name at hotels), resigned all his offices in his native land and came abroad to rally from a combination of bronchitis and nervous prostration, which, various doctors say, might have been expected in the (altered) circumstances. With this aspersion we have nothing to do. Still, as one who has spent more than thirty years in church-and-charity work, I may express the doubt whether the friction of pastorate or professorship could have worn more seriously upon the nervous system of the average man than the continual conflict of opinion and belief which goes on between his spouse and himself. But Dr. Sharpe is many removes from the average man.

They have left their carriage and stand below the projecting raiment of a round tower upon the outer wall. Close to the ground are sections of ancient Roman masonry; the tower is topped by a modern shed. The outlines of a window, almost wholly blocked up with rubbish, appear below the rude building on the summit.

"That," says our guide, "tradition points out as the window from which Paul was let down in a basket, as recorded in Acts, the ninth chapter and twenty-fifth verse."

I cannot record that Dr. Sharpe pricks up his ears as the innocent remark enters them, for they are already red and bristling with the heat of argument. He turns energetically upon us.

"I do hope," he says, "that as intelligent Christians of the nineteenth century, you will not credit this worthless—utterly worthless—tale concocted by monkish charlatans. Setting aside the extreme improbability that the dwellers

POSTMAN ACROSS THE DESERT FROM BAGDAD TO DAMASCUS.

in the Damascus of that day would have regarded the spot as noteworthy, the masonry of the wall belongs so evidently to a later period that one is lost in surprise at the impudence of the attempt to identify the locality with the manner of the apostle's escape from his persecutors. And cannot you see, my good friends"—haranguing the guides of his party and of ours—"that you are guilty of folly, if not of mischief, in perpetuating the legend? Not that I blame you as much as I blame the superstitious ignorance of the travelers who encourage you to repeat such stuff."

Dr. Sharpe is tall and meagre, with a long face running down into a pointed red beard. He waves his arms like a windmill as he declaims. His wife is much below medium height and plump as a quail, with bright, beady eyes not unlike those of the same bird; her voice is a mild twitter, but we feel at once that for "staying power" she can be counted upon as no mean antagonist for her irascible husband. It may be the staying quality of the feather-bed that yields only to rebound, but it is "all there."

"My love!" is the preface to her remonstrance—"where *would* the Church be now, but for these traditions? When I see the unhappy effect produced upon *your* mind and spirits by incredulity, I am disposed to cling *more* closely to what holy men have transmitted to the generations following. For instance, you took no comfort *whatever* in our visit to the scene of St. Paul's conversion—and I had *such* a precious time!"

"No comfort! Where should I find satisfaction in a baseless legend, fabricated originally, I doubt not, on purpose to induce travelers to stop and buy refreshments at what, it is pretended, was an ancient resting-place of caravans from Jerusalem to Damascus. My good wife—" turning to us, as he says this, an audience being a vital necessity with men of his calibre and class—"my good wife formed her belief upon this, as upon other subjects of sacred history, by the help of pictures and goody-goody story-books. She has in her mind every detail of the scene as depicted by these; Saul—he wasn't even Paul then, my dear, much less a saint—tumbling from his horse upon a macadamized road, and his companions in various attitudes of a like plight, when it is not certain, or even probable, that he was ever upon a horse in his life. It is more likely that he made the journey on foot, or upon a camel, or an ass. As to the location of the event, I dare say that each of you guides"—in sudden appeal—"has heard one or more places named besides Katana, as the scene of Paul's conversion?"

The men look at one another, and even grave David Jamal smiles as he answers that "Katana, on the ancient route to Jerusalem, has been for centuries pointed out as the place where the Apostle Paul was struck to the earth and had a direct revelation from Christ, our Lord. Other places have been suggested, but the best authorities agree upon Katana."

I think there is something in the Christian dragoman's way of pronouncing the words, "Christ, our Lord," that tends to silence the disputants. At any rate, they do not resume the discussion until we are out of hearing.

"I have heard of that gentleman before," observes David, as we drive away. "He is, I believe, very learned, and is collecting materials for a book upon the Holy Land. Up to this time, it is said, he has found nothing anywhere that he can believe in. His book will not be large."

"PAUL'S WINDOW" IN DAMASCUS WALL.

Encouraged by our laugh, he adds: "I recollect his wife as one of a party of thirteen ladies I took through Palestine over twelve years ago. Miss De Credo was her name. I should have said, then, that she would have been a Roman Catholic by now. She set great store by relics and the like."

We halt to examine another bit of city-wall, almost ruinous above, but resting upon a sure foundation of Roman architecture. In spring-time, when the encompassing orchards are in bloom, Damascus may deserve, in the mind of the approaching traveler, some of the encomiums lavished upon it by poets and historians of a former age. At this season, the "pearl" is undeniably dingy. No modern Mohammed will pause without any one of the four gates to compare the attractions of this, as an earthly Paradise, with those of the heavenly, much less turn away with a sigh of renunciation, saying: "Man can have but one Paradise and mine must not be here!"

"Certainly not, if Damascus odors were the same then as now!" remarks my traveling companion.

Diverging from the road skirting the city-wall are others leading between walled orchards into the outlying country, some stretching far away to cities whose names kindle wistful gleams in our eyes. Upon that leading to Bagdad—

"ANOTHER BIT OF CITY WALL."

distant, from thirty-five to forty days' journey, as a caravan travels, ten by the "swift dromedaries" that carry the mails—has halted a long string of camels. We check the driver to have a better look at the ugly and at present patient beasts, laden with who can say what wonders of that far and fabled mart? The camel is notoriously uncertain of temper, captious, irritable, and sometimes so sullen as to refuse food for a fortnight at a time, during which period he must be muzzled to prevent him from biting his fellows and drivers. We hear of another peculiarity while we survey this group.

"They are of the country and the desert," we are told, "and must be unloaded here, without the gates. Not being used to city streets and sights, they would be unmanageable if taken into the town."

There they stand, with the outstretching necks that look at once supple and stiff, meek and stubborn, awaiting the command to let themselves down groaningly upon their much-doubled-up knees that their burdens may be unstrapped—fascinating studies, if only because of their irredeemable hideousness and the

"OF THE COUNTRY AND THE DESERT."

physical idiosyncracies that set them apart from other quadrupedal domesticated creatures. I always stop to look at a camel in passing, and never without a shudder of wonder that is not admiration, and loathing which is antipathy.

Passing through the city-gate into the street which is called Straight and, with the memory of the Rev. Zwinglius fresh in our thoughts, negativing the jesting proposition to visit, first the house of Ananias, and then the grave of Cain—this latter being a shapeless pile of stones cast together by the passers-by in execration of the first murderer—we alight presently at the entrance of the world-famous bazaars of Damascus. Hard-by yawns a mighty, blackened shell; the walls still standing and the roof arched above them, but through gaps in the rude boarding filling up the doors, and through the empty window-frames, one sees what devastation has been wrought, and recently, by fire. The bazaars nearest to the ruin have not escaped the licking flames that burst from the windows. A stately minaret towers heavenward near to what was the great mosque of Damascus.

"There was nothing in Syria like it," relates our guide. "Men came many miles—oh, very many—to pray here. They believed one prayer in this mosque to be worth thousands and tens of thousands said in another. And that God would leave this holy place standing for forty years after the world should be burned up, in order that those who had not yet found the Gate of Paradise might have a place for prayer."

"How did it take fire?"

The two dragomen exchange looks. One steps near and lowers his voice:

"From the melting solder with which workmen were repairing the roof, say the Christians. Others"—avoiding in the crowd the mention of Mohammedans or Moslems—"will have it that the fire fell from heaven upon the Mosque because Christians were employed in building it after it was burned in part many years ago."

We have just fairly entered one of the principal streets of the bazaars—a continuation of that "which is called Straight"—when a roar as the sound of many—and musical—waters, draws our eyes to the tossing crowd filling the lower end. The guides have barely time to draw us within the shelter of a doorway when the human tide rolls up to our feet, choking the narrow thoroughfare—a hustling, struggling mass. Foremost are perhaps a hundred boys, clapping their hands high above their heads and screeching, rather than yelling, what David translates in our ears as "Praise to the Prophet!" A herd of twice as many men is upon their heels, vociferating the same words, shaking tambourines high in the air and beating small drums while whirling them aloft.

"What is it? A popular revolution?" we gasp, as the billows roll away in the distance, and the air settles about us in comparative quiet.

Both guides smile, evidently in no wise excited by the demonstration that took our breath and well-nigh our senses away.

"They are men and boys who have offered their sacrifices, without pay, to the government to build again the Great Mosque," replies a guide. "They go thus about the streets crying, 'Praise to the Prophet!' that others may be moved to do likewise. We hear, to-day, that the Sultan will give generously, some say, more than one hundred thousand mejidie"—the Turkish dollar—"from his own purse toward the great work."

A pursy Englishman, who fairly exhales his nationality, so obtrusive are his round-crowned hat, mutton-chop whiskers, red face and tweed suit, is crowded against us, and has the benefit of our explanation.

"Rayly!" he utters, coolly. "Now, do you know, I took it to be a sort of Eastern Salvation Army, don't you know? The tambourines, drums, and all, look like it."

The roar is not even a whisper by this time. Besides ourselves, and our Briton, nobody seems at all interested in the doings of the clamorous crowd. The fine-looking Turk, sitting cross-legged upon the counter of his stall over the way,

"THE STREET THAT IS CALLED STRAIGHT."

has dropped the lids upon his dark eyes with as fine an expression of boredom as could be achieved by an English exquisite at the height, or depth, of a London season. At the stall to his right, a customer is beating down the price of a narghileh (or water-pipe), with an amber mouthpiece, with no more apparent consciousness of the passage of the mob than of our observation of himself and his actions.

"This is the Oriental calmness of which we have heard, all our lives," we soliloquize, "and these narrow, dirty lanes are the great bazaars!

"The poorer bazaars, of course?" we pronounce, in cheerful certainty of the reply.

The Damascus guide does not catch our meaning. The question is put in a different form:

"There are broader streets and finer shops in Damascus than these?"

To prepare ourselves for this expedition, we have been reading a chapter in "Mirage," a strangely sad story published perhaps a score of years ago, and a sentence is freshly distinct in our minds:

"They were deep in the bazaars of Damascus—deep in those cool, dim, vaulted spaces, with the lustre of silks, the gleam of metal and porcelain about them, and the odor of gums and curious spices filling the dusty air."

Certainly the air is dusty, after the rush of the not-clean mob described in a former paper, and there is an odor—there are more than a score of odors—that fill the vaulted space. If there be one sense which the traveler in the East comes to regard as superfluous and inconvenient, it is that of smell. If I refer repeatedly to this fact, it is because the pressure of the truth is never absent from us. Being told that the scene before us is a favorable specimen of what we are here to see, we resign ourselves to the Oriental odor, "curious" to the uninitiated, but freighted with no spice or gum with which our olfactories are acquainted, and set our eyes to work instead.

We are in a network of narrow streets, none of which are twenty-five feet in width. Low, arched roofs cover them; the light breaking, here and there, through slits in the arch, in a deliberate, dreamy fashion essentially Oriental. As the sight becomes accustomed to it, we see that the glints brought out by the intermittent rays come from stalls piled and hung with silverware of every description.

"The street of the silversmiths," Abraham observes. "The trade is in the hands of the Christians. The man over there has the same shop which his grandfather kept. The Christians, in former times, could not hold such property as cattle, houses and orchards, for it was often taken from them or destroyed by—those who were of a different race."

(Again we notice the scrupulous avoidance of the word "Mohammedan," or "Moslem," in a public place, and speculate as to whether prudence or courtesy dictates the custom.)

"Therefore, Christians usually are in trade—manufacturers, and like that."

What are dignified by the name of shops are really stalls, such as line our markets at home, but enclosed on three sides. A movable door, a sort of wooden shutter—sometimes in one piece, sometimes in several—closes the front at night. When this is taken down, the shop is opened without other ceremony. A big window of one of our great dry goods stores would hold the entire stock-in-trade of any of these recesses, and leave room to spare. Upon the counter, or upon a divan behind it, the proprietor has tucked his slippered feet away in the folds of the baggy cloth trousers which, I insist, are the reverse of graceful, and pulls slowly at the yards of flexible tube connecting the amber mouthpiece between his lips with a vessel exactly like the carafe, or water-bottle, used upon our dinner-tables, with the exception of the tobacco-pipe attachment.

We are collecting—with the rest of the traveling "foreigners"—souvenir spoons, and mean to procure some in Damascus. The expression of the intention is all we have to do for the present. David steps to the front as the coveted articles are produced. The prospective vender is an acquaintance, nay more, a friend, for they kissed one another just now, and the Damascan addresses him as "Abou-Nassar," the "Father of Nassar." When a man becomes the proud owner of a son, he is thenceforth known among his intimates as that boy's father. Ten daughters may have preceded the youngster, and six brothers may follow him, but the parent continues to rejoice in the title of father of the first-born son.

These are two friends, as I have said, but as the buyer's representative, the dragoman is transformed forthwith, until the sale is concluded, into the bitter enemy of his countryman.

"Never give the first price named!" is the first lesson taught the foreign visitor. The sum may sound reasonable to us, but a cardinal principle of Eastern shopping is to believe and hold for certain that it is iniquitous, and the universal practice is to beat it down to at least one-third less. It matters nothing to our honest representative, who would not cheat his worst enemy of a penny, that the amount involved is only a franc, or, at the most, two. Figuratively, he girds up his loins, and rolls up his sleeves, and "pitches in" for a wordy duel. The voices of both combatants rise, their gestures threaten each other's lives; one-tenth of the noise in an American shop would call in the police in force.

Quite aware that it is all sound, and no fury, we lean back in the rush-bottomed chairs set outside for us in the open street—for sidewalks there are none—and give ourselves up to contemplation of the vistas stretching away from us on four sides.

After the shadeless glare of the Damascus sunshine, we recognize the cool half-lights as grateful, and the occasional streams of radiance through the apertures in the roof bring out picturesque features in merchandise and costumes. At the far end of the street are shops wherein are manufactured the various stuffs here offered for sale.

One street is given up wholly to the display of rugs; another to a bewildering succession of rings, curios, brass trays, cups, bowls and the like; another to mother-of-pearl tables of divers sizes, patterns and heights. A principal avenue exhibits on both sides "Franji" articles. The manufactured term, in universal use in this country, and referable, I think, to no known language, covers whatever smacks of European or other Occidental usage, and is freely applied to all sorts of innovations. It signifies, in the present instance, French, English German, Italian, and even American goods—calicoes from Manchester, tinware and cutlery (cheap) from the United States, colored glass from Baden-Baden and every variety of knick-knacks from Paris. Several streets away from where we are awaiting the result of the pitched battle of Arabic tongues are the meaner quarters devoted to the sale of second-hand clothing—it is needless to say by whose hands.

"IT IS NEEDLESS TO SAY BY WHOSE HANDS."

It may gratify their American sisters to know that those whose lot seems to us hard enough without such a barbarous restriction are not debarred from doing their own shopping in Damascus.

Old as is the beclouded "Pearl of the East," the "inferior sex" has the advantage in this particular over Nazareth, Nablous, Gaza, Jenen and Jaffa, where native women, veiled or unveiled, are never allowed to visit shop or market-place. Even the Christian matrons must apply to this occupation, so dear to the feminine heart, the injunction of the Great Bachelor (or widower?) of Tarsus, and ask their husbands at home concerning the latest sweet thing in jewelry, and the "Franji" gown-patterns put up in decorated boxes in Paris, and sent eastward at the end of the season for the delectation of hareems and Greek, or Roman Catholic, or Protestant households.

How do they support existence in such circumstances?

Fathers, brothers and husbands, in unconscious prevision of Mr. Edward Bellamy's "Looking Backward," select and bring home samples of cloth, silk, etc., and the secluded wife, daughter or sister chooses from among these such stuffs as she imagines, from the bits laid before her, would be handsome and becoming. Those who know how unlike the sample is to the effect of the fabric when seen "in the piece," especially as displayed by the cunning hands of a trained salesman, can guess how frequent and sore are the disappointments when the purchases are unfolded in the women's apartments.

That tall woman, whose mufflings cannot conceal the outlines of a figure uncommonly good for a land where corsets are unknown, and a "neat fit" is

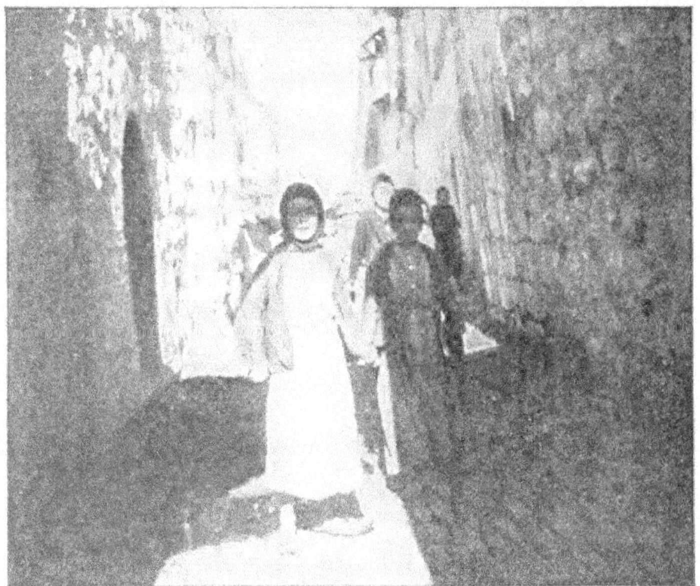

"THEY MARCH MORE SOBERLY NOW."

almost as rare, makes as serious a business of her bargaining as if she had the interests of a brace of defenceless "foreigners" upon her conscience. She stands all the while, we note. The uncompromising rush-bottomed chair would seem to be reserved for foreigners. She haggles over a pair of silver filagree bracelets, fitting them upon a round, brown wrist, and slipping them excitedly up and down, chattering volubly in Arabic in a voice so dissonant that her "mendeel" shivers, rather than wavers, before the torrent. It is dark blue, with red poppies, larger than life, "powdered" over it. All that we see of her face is the indication (to use a technical phrase) of two big, black eyes in the upper part. She can discern clearly not only the countenance of her natural enemy, the tradesman, but defects in the workmanship of the trinkets she is trying on, for she calls his attention sharply to a misshapen link, tears them from her wrists, tosses them disdainfully upon the counter, gathers the voluminous "izzar" about her and turns away with the air of one resolved to endure no more imposition. She has taken two or three steps when the merchant, picking up the discarded ornament, says something quietly and indifferently. As indifferently she pauses and answers him over her shoulder.

A minute later she is again at the counter; the salesman is folding the bracelets in tissue-paper, and she—still as if she would rather leave than take them—is extracting from her purse the sum finally agreed upon between them.

SHEARING A CAMEL.

The affair is a farce throughout, and is repeated often enough to lose all novelty, not to say charm, were the disposition to make a bargain less obstinate in the human breast. The buyer knows perfectly well that the "asking-price"—a term with which one speedily becomes familiar over here—is as much above the real value of the article offered as what he, on his part, proposes to give for it is below it; he is also certain that somewhere between the two will be the final point of settlement.

The responsive yawn is arrested midway by David's abrupt termination of hostilities. He has pushed aside the spoons, and, generous indignation in every line of his visage, motions us to follow him.

"But"—expostulates one of the unsophisticated—"since they are what we want, and not really dear, would it not be better to pay his price?"

"Like the rest of them"—enunciates the dragoman, judicially, without specifying who are included in the pronoun—"he thinks travelers are made of money. You will get the spoons. He will call us back."

The spoons are in my trunk, and are probably worth a half-franc apiece less than was paid for them after the profuse expenditure of breath, nervous force and minutes. Shopping, always wearisome work to busy people, gathers new horrors from such accessories. By the time we have halted here fifteen minutes, and there twenty, and somewhere else ten, we are thankful to be out of the covered arcades, and to see the unobstructed blue overhead in a street where we meet one wing of the boyish troop who passed us an hour ago. They march more soberly now, but still chant the sing-song, "Praise the Prophet!"

The sun is low and red upon the horizon as we, for the second time, drive in at the eastern gate of the aged city. A single camel is left in the open space without, where we saw the caravan at noon. He lies so prone and abject upon the earth, his neck stretched out to a length so incredible, that we are divided between the desire to pity and to deride him. His acknowledged master stands above him brandishing a pair of shears; one side of the forlorn beast is already so naked as to put Sterne's shorn lamb to the blush, and as the crimson sunlight strikes athwart him, the "stranded ship of the desert" looks abnormally and inconceivably hideous—even for a country camel.

CHAPTER XII.

BY THE SEA OF GALILEE.

"HER FAT PUPPY."

AWAY from Bethsaida, the city of Andrew, James, John, Peter and Philip, we took a handful of curious sheils, as black as coal. The beach is strewed with them, and they are so unlike the gray and brown we find at other points of the sea of Galilee that we cherish fanciful associations connecting them with the silent desolation pervading this coast. "City," there is none. A few half-buried stones mark the site of the town built by the Philip who married Salome a little while after she danced off the head of John the Baptist. He was the most respectable member of the Herod tribe, we recall in picking up our shells, but we look in vain for the handsome tomb built here for him, about the time that the Cross which was to draw all nations unto it was lifted up on Calvary.

Capernaum lies a short hour's ride from Bethsaida. The horses pick their way over the pebbly beds of what will be torrents in a couple of months, and eye longingly patches of coarse grass, greening where the stones are not so thick; we pass a ruined Roman aqueduct used in the days of the Herods for conveying water to the fine cities they built upon the sea called by their Emperor's name. In the immediate vicinity of Capernaum, reeds and flags must have hidden the fallen stones last spring and summer, for they rustle drearily in the wind moaning up from the sea, and tufts of dried herbage define the outline of such fragments of limestone as have not been utilized in building the Roman Catholic convent we see rising up in the background. It is unfinished, but there are no workmen to be seen to-day. By the time we have alighted, the entire population of Capernaum is drawn up in hospitable order to welcome us. Pitched where, for aught we can say to disprove it, Matthew's custom-house may have stood, is a small black tent, and from the smoky recesses emerge a Bedouin, as venerable and benign as Abraham; his wife, throwing a fold of her dark blue nondescript garment modestly over her chin as she comes; a daughter of fifteen, or thereabouts, bashful where her mother is modest, and the prettiest native child we have seen, a girl of four, whose conviction that we will covet and probably steal her fat puppy, overcomes curiosity during our stay.

The patriarch is the care-taker of the sacred precincts. The smile with which we hear his office named is exchanged for disappointed looks upon discovering that what we especially desired to see is nowhere to be found.

"Yet it was certainly here last year," David says, his soul divided between surprise, sorrow and wrath. "If I am not mistaken, Dr. Geikie speaks of it in his book, and Dr. Talmage will have told you how impressed he was by it when he was in Capernaum, three years ago. It is more than pity that some of those thieves who are all the while hanging about such places have stolen it. I say it!

"THE ENTIRE POPULATION OF CAPERNAUM."

even if it has been built into those walls over there"—nodding ominously in the direction of the unfinished convent.

After a moment of painful silence, he resumes: "You see, it helped to make it all so plain! I have stood here, many times, and imagined just how He looked when He said it. It was very interesting to think that His eyes must have rested upon it often, and that He may have pointed to it when He said, 'Our fathers did eat manna in the desert'—as you will read in John, sixth chapter, thirty-first verse, and again in the verse forty-ninth, 'Your fathers did eat manna in the wilderness, and are dead. This is the bread which came down from heaven'—meaning

MOUNT OF BEATITUDES.

Himself. It is all about the manna and the living bread, you see. Who can doubt what put it into the mind of Christ, our Lord, to preach upon that subject in or before the synagogue that used to stand hereabouts!"

We stroll up and down wistfully, musing, while the population of Our Lord's "own city" watch us, mystified, but silent. Nobody explains to them why we have come or why we are dissatisfied. A query in Arabic has proved that they know nothing of the disappearance of the massive block of limestone, probably part of the façade of the noble synagogue now leveled to the earth. Upon it was a bas-relief of the pot of manna kept within the Ark, and afterward in the First Temple. It would have been easy, in looking upon it, to reproduce in fancy the scene referred to by our guide.

"IT WAS CERTAINLY HERE LAST YEAR."

"The Bible is the best Baedeker for Palestine," says Alcides, for the hundredth time, and we seek for ourselves seats upon the fallen stones to read up in our invaluable "International Teachers' Edition," the passage indicated under the head "Capernaum."

One cries out presently, "Why may not this synagogue have been that built by the centurion who loved the Jewish nation?"

It is not unlikely, since the bits of carving from which we proceed to clear the dead grass show skill and taste. The curling acanthus leaf of a Corinthian capital is embedded in a tough tussock, and there is a richly sculptured corner that may have belonged to a frieze.

"And this is the Capernaum, once exalted unto heaven by these and many other wonderful works—*now!*"

We gaze with new comprehension of the story and the prophecy, upon the grass-grown stones, the black tent that makes the picture more, not less, desolate, and in the background, the useless convent, where able-bodied men will patter prayers in an unknown tongue, and exalt above the Divine Son, the very human mother who broke in upon the most solemn sermon that has come down to us from His lips, by the request that He would shake off the crowd hanging upon the inspired words, and come out to speak to her. "His own city," with none to take pleasure in her stones, or to favor the dust thereof!

The folded tents, the camp-equipage and mules have gone forward toward Tiberias when we return to Bethsaida, but David and Alcides alight to survey the ground and make sure that nothing has been overlooked, consigning Dervish and Massoud to the care of Serkeese. In consequence of which misplaced confidence we have a diversion from the sober reverie that has held us since we left Capernaum.

Massoud is not a colt in years. He will never be otherwise than coltish in heart. There is a clatter of hoofs among the irregular boulders and pebbles, a jocund neigh, and Serkeese stares open-mouthed after the fleeting steed, letting fall Dervish's bridle in the excess of his astonishment. No fences obstruct the gambols of the merry beast, who would have a companion in his scamper were Dervish as light of heart and heels as himself. For one mortal hour he leads three bipeds a dance up and down the beach which none of them can forget. At every hundred yards or so, he stops to laugh in the perspiring faces of the pursuers, waiting until a hand is almost upon his bridle before he throws up head and hoofs together and trots gently out of reach. Viciousness is as far removed from the heart of the little horse as the intention to surrender himself voluntarily to any of the trio. Each mentally recalls the story of a similar escapade upon the plain of Esdrælon when the tricky creature eluded capture for twelve hours, keeping the whole party waiting upon his pleasure.

He blunders at length, when even the solitary looker-on appreciates that there is a time to laugh, and a time to refrain from laughing; takes a wrong turn and is followed by David into a little pass forming what is, in sporting phrase, "a pocket."

"He cannot do justice to the subject!" quotes Alcides, aside, as, having seen his passenger again in the saddle, the dragoman receives Dervish from the hand of the abashed cause of the misadventure, with never a syllable of comment.

"A fool uttereth all his mind; but a wise man keepeth it in until afterward."

This particular wise man has something to say by and by, and, as usual, something pertinent.

"We have lost time, and shall be caught in the rain before we can get into camp."

By the time we draw rein upon the slope—green even in winter—where tradition says our Lord commanded the five thousand men, women and children to

"MISPLACED CONFIDENCE."

sit down upon the grass, and fed them by the hands of His disciples with the miraculously-multiplied loaves and fishes, the sky is one cloud, dark-gray and sullen, the lake upon our left a gloomier reflection of the pall above it, and stirring uneasily in an occasional flaw of wind.

The use of the word "sea" misleads many as to the size of this, the most celebrated sheet of water upon the globe. I is, in fact, but fifteen miles in length and half as wide, and, apart from storied interest, as unremarkable as scores of mountain lakes in America whose names are not known beyond the special townships that border them.

It is here, as the surface begins to be pitted with large, slow drops of rain, that we listen to a true tale confirmatory of the Scriptural account of the storms which overtook the "little ship" in which our Lord slept, his head upon a pillow in the boat which the toilsome rowing of the disciples could not bring to land, until, in the fourth watch of the night, the pressure of His feet stilled the billows, and His voice rebuked the "contrary" wind.

But a year ago a tempest rushed down Lake Gennesaret so fierce and sudden as to tear from their foundations and wash into the sea thirty houses from the town of Tiberias. Now, as of old, the rage of wind and wave is intense and short-lived.

Tiberias can never be comely. Seen under a dripping sky, it is repulsive. The alleys that do duty as streets are lined, in some quarters, with booths or

THE CAMP.

stalls, in others with mean dwellings, all dirty-white in the sunshine, dirtier gray when wet. The inhabitants have retreated to the rear of the shops, even where the streets are covered with matting or with leaky wooden roofs, and eye us listlessly as we pass, the horses plash through filth made liquid by the rain, and coursing in vile smelling rivulets down the middle of the street. The situation of the town, and the general noisomeness remind one of Jaffa, although Tiberias is much the smaller of the two.

Pleasant exceedingly is it to arrive at the camp pitched beyond the outskirting houses and so near the edge of the lake that the wash of the water upon the banks promises a lullaby throughout the dark hours. By the time luncheon is disposed of, the rain abates, and we sally forth for a walk upon the wet stones.

The custom of washing the bodies of the dead in the Sea of Tiberias is of immemorial antiquity, and was, until recently, almost a daily occurrence. Corpses were brought from the surrounding country and across the lake, and committed to a sect whose profession it is to prepare them for burial by these ablutions. The details of the operation are too disgusting for publication, the process of evisceration being at once ingenious and revolting. Not until the water pumped into the body leaves it as clear as when it entered, is the work considered complete. As described

"ONCE EXALTED UNTO HEAVEN—NOW!"

to us, the business is unpleasantly like that of preparing a fowl for cooking, even to breaking certain small bones of the limbs. Our informant, a resident of the town, although a foreigner and a Christian, adds the gratifying intelligence that a strenuous movement is on foot to break up the practice altogether, and that such success has followed it as to intimidate those who have made it a means of livelihood.

The world moves—even in Syria.

The sun sets gloriously over the lake, some peculiar effect of refraction bringing the opposite shore nearer and nearer as the amber glow deepens. We look out upon a sea of glass mingled with fire: the white tents are washed with yellow light; there is not breeze enough to lift the American flag floating from the peak of the largest of the group, and we see in tremulous yet distinct outline, to our right across the water, the roof-shaped "steep place" down which the herd rushed violently into now placid waves.

Fish, fresh from the lake, follow the soup at dinner. The water teems with them, as when the weight of an hundred-fifty-and-three strained, without breaking, a mesh of the net cast at the Lord's command upon the right side of the ship. We talk, over the savory mess, of the story that the hook sometimes brings up one oddly marked with a black spot upon the head and under the throat in commemoration of the grasp of Peter's thumb and finger upon the prize that yielded up the piece of money.

"A 'stater,'—in value two shillings and a sixpence," we gather from our "Palestine guide book," and consult dragoman and John-of-the-many-names, as to the authenticity of the tale.

Neither of them has ever seen such a fish. David, however, has heard of it before from Mrs. Sharpe.

"When she was Miss De Credo, of course," he subjoins, with no intention of being satirical. "She quite believed in it."

After dinner we again seek the beach. The new moon sinks below the low hills of the horizon; the swish of the wavelets blends musically with the low hum of talk from the camp behind us.

> And the sheen of their spears was like stars on the sea,
> When the blue waves roll nightly on dark Galilee,

repeats a dreamy voice. "Byron never saw Galilee. But the stars and the sheen are here."

THE TOWN OF NAZARETH, AS IT APPEARS TO-DAY.

CHAPTER XIII.

WHERE HE WAS BROUGHT UP.

WE have never been so nearly in sympathy with Dr. Sharpe, or so intolerant with his wife's devout credulity, as after visiting the Latin and Greek churches built above the spots where the Virgin Mary received the visit of Gabriel and the announcement of the great glory to be bestowed upon her.

Each is in a crypt, and going first to that of the Greek church, we look respectfully, if not reverently, at the dark, damp chamber, supplied with altar and the usual paraphernalia of the underground or the upper-world shrine—candles, artificial flowers and other tawdry ornaments. We are in a sort of cellar, the stone walls of which are grim with time. A well yawns in the floor; beyond it is a locked gate. The sacristan thrusts his arm between the bars, and throws the light of a torch upon a flight of steps leading up into the darkness.

Our guide mechanically translates the Arabic that gurgles from his tongue:

"This church stands upon the site of the house of the Blessed Virgin. This was her kitchen. Here she was busy at work when the angel appeared to her and said, 'Hail, Mary! thou that art highly favored, for thou hast found favor with God.' The fountain burst forth from the rock in celebration of the event."

A cup is lowered into the well; we drink thereof, cross the palm of the sacristan with a piece of Turkish silver, and mount to the surface of the earth, obediently following our leader to another church, where the same story is told, almost word for word, of a second subterranean chamber and spring.

A little heart-sick and a good deal disgusted, we thread the public ways of the "white city," discouragingly new in appearance, and long, as we have often and elsewhere longed, for something that will restore the lost sympathy with what the place should recall.

"Mary and Joseph lived here," we say to ourselves and one another. "Here was the scene of the Annunciation; hence Mary went into the hill country in haste—perhaps up that slope over there—to visit Elizabeth. Here, her Son grew in favor with God and man. Here, cut to the quick by the home truths spoken to them by Him in later years, the townspeople with whom He had been so popular as a youth hurried Him to the brow of the hill on which the city is built, and would have ended His ministry then and there but for a miracle."

All this we know and believe, but cannot feel.

At a turn of the street we are confronted with something that alters the current of discontented meditation. A lad stands just without the threshold of a carpenter's shop. His tunic of blue serge is girt at the waist with a red sash; red slippers are upon the unstockinged feet; the red tarboosh, or fez, worn by all Turkish subjects, sits jauntily upon short, fair curls. At sight of the leveled kodak, he laughs, and his hands chafe one another nervously. From the recesses of the shop, black as midnight when one looks in from the outer light made more glaring by reflection from the white houses, a voice calls in not unkindly admonition; the boy laughs again, brightly and fearlessly, in moving away to his work. A bench extends from the door to the rear of the windowless shop, and two men, one turbaned, the other wearing the tarboosh, are busy about it. Our lad drops flat upon the ground, crosses his legs under him to gouge a square hole in a block into which a beam is to be mortised. He handles the chisel deftly, and, now that he is at it, so diligently that he does not give us another look. But for the modern bench, the picture must be very like that which has been a theme belovéd of such artists as Holman Hunt and the earlier masters. Presently, the elder of the men, who stands near the door, glances around, his notice attracted by our shadows upon the floor, and touches, first his forehead, then his heart, bowing profoundly:

"HE LAUGHS AND HIS HANDS CHAFE ONE ANOTHER."

"*Naharak saïd!*" (Peace be unto you) he says gravely.

"And unto you!" we instruct our interpreter to reply. The carpenter is an old man, but hale, with clear eyes and a full white beard.

Not a bad model for a Joseph in a "Holy Family," we agree, after watching him for a minute longer, taking in all the accessories of the tableau, even to the play of a stray sunbeam upon the yellow shavings curling from the bench to the floor.

There is fascination in the scene, and soothing. We are brought again into touch with a past that filled our minds at the first sight of the hillside town. Thus might the supposed father and docile Son, "subject" to all the conditions of a Galilean artisan's life, have toiled together for many a day and week during the thirty years of preparation for His ministry and His death.

"*Mar salaami!*" (Good-bye)—we call out at length, moving on, and between the deeper tones of the two men rises the silvery treble of the apprentice:

"*Mar salaami !*"

"In what language did our Saviour speak?" questions Alcides, suddenly.

The query is apparently new to David Jamal and our other escort, a Nazarene Christian and a graduate of Beirût College, a fine, intelligent young fellow, by the way, and an honor to his Alma Mater. It seems quite certain, after some minutes' talk, that in a country where customs have changed so little in nineteen centuries that one can hardly take a dozen steps without seeing or hearing something illustrative of Biblical history—the peasantry at least may use the dialect familiar to the household of Joseph, the carpenter. Did the musical phrase of farewell that lingers in our ear often pass *His* tongue?

We are still talking of it when another turn brings us into an open square that must be the rallying-point for both sexes of all classes, so alive is it with movement and murmur. Right before us is a group of children, chiefly boys, of varying ages, arrested in some game by the abrupt appearance of strangers. The foremost and tallest boy lays a reassuring hand about the neck of a mere baby who is his charge, and blinks inquisitively at us in the broad sunshine; in the background on his right hand stands a much smaller child with a face of singular beauty. It is not possible to regard the children, especially the boys of Nazareth, with indifference, and there is something in this encounter that dizzies one by the swift rush of association.

"What do they know of the child Jesus?" we interrogate the young Nazarene student.

"Those who have Christian parents think and speak of Him as do the children in your land and other Christian countries. The rest know little of Him and care nothing for Him. We have here Sunday and day schools in which the Bible is taught, and the story of our Lord's life and death is well known to the members of the Latin and Greek churches."

More children! The soft thud of their bare feet upon the pavement is hushed as they halt and huddle together in the exact focus of the insatiable kodak—neither timid nor bold, but moved by a common desire to inspect and make acquaintance with us. They are healthy, well-formed and evidently well cared for. In all Nazareth, as we note with gratification, we see not one deformed child, or one professional beggar of tender years. All this may signify less than we fancy, but, without quite defining the sentiment to ourselves, we are glad that these things are so.

"The fountain of the Virgin," says our guide, and it is not in tourist-nature to withhold the exclamation—"How many more?" We change our tone at a second glance. No monkish fable of kitchen and miraculous gush of living streams clings about the one perennial well in a district mainly dependent upon rain-cisterns for a supply of water. For hundreds and hundreds of years this fountain

"A GROUP OF CHILDREN."

has borne the name by which the world knows it, and even Dr. Sharpe cannot deny the probability that it was here when Mary kept her husband's house in Nazareth, and her Boy ran by her side in street and market-place, or clung shyly to her skirts, as that pretty toddler across the square steadies his trial steps by holding with both hands to his mother's izzar. The water leaps from a spout in the face of the wall into a stone trough, escaping thence into a tank. An arch is built above it, and steps lead down to tank and trough. These steps are crowded with women, some rubbing out, and some rinsing clothes as composedly as if the open archway were a back laundry at home; others bearing down empty jars in their hands to be filled at the spout, or trudging up the wet stairs with full vessels upon their heads. They chatter cheerily among themselves, and the most notable are in no hurry to hie them back to home and indoor labor. Of all marketable possessions, time is least considered in the East.

Sometimes, as we are not surprised to hear, there are wrangles over the common wash-tub, and quarrels as to precedence in the matter of filling the great

"THE FOUNTAIN OF THE VIRGIN."

earthen jars. In the sunny noon of to-day good humor prevails, and sociability is at high tide.

Mary was a peasant and working-woman like these. We check ourselves, as if guilty of irreverence, in wondering if she bore the narrow-necked "pitcher" upon her head with the audacious grace of that laughing girl who does not touch it with her hand, and if she adjusted it, before mounting the steps, upon a many-folded cloth such as the handsome woman down there has interposed between her veiled head and the dripping vessel. Yet—and again—the highly-favored among women was a daughter of the people, and her Son, at thirty-three years of age, had not where to lay his head.

A rhythmic murmur hums up the street leading to the open space about the Virgin's Fountain; the women cease their gossip; the children run off in the direction of the sound, and we follow more leisurely. All recognize the tuneless chant with which the mourners go about the streets. It is a Latin funeral, and the procession is led by a fat priest in full canonicals. An acolyte, clad in what looks like a white dressing-sacque, swings a censer at his side. The coffin, borne upon the shoulders of friends or neighbors, is of a vivid scarlet with white bars crossing it. Behind a little company of what we take to be relatives and acquaintances, six

MODERN NAZARETH.

cowled friars walk two-and-two. They are stalwart of frame and bearded up to the eyes; upon their large feet are strapped leathern sandals. Not a woman is seen in the band of mourners or among the lookers-on that straggle after the bier into the church.

We remain without the burial-ground until the short service is over. Church and cemetery are on the slope of the hill which rises far behind them. The whole town has the air of holding on for its life to the declivity. Every house is literally founded upon a rock, and the circumstance is but one of the numberless proofs

CANA OF GALILEE.

daily laid before our eyes of the care with which the Nazarene Teacher studied the world in which He was brought up, and His tact in presenting to the "common people" analogies and types drawn from the life He shared with them.

The grave is partly dug, partly hewn in the stony ground, and the bright red coffin, containing, we are told, the body of a young man, is set down upon the pile thrown up beside the mouth of the pit. The throng, augmented by many who have gathered from all quarters, closes about it, and the prayer committing dust to dust is droned above it. Until now, all has been conducted as quietly as in an English churchyard, but at the conclusion of the service, the father, who has stood at the head of the grave, motionless, with folded arms and averted face, casts himself with an "exceeding loud and bitter cry," upon the coffin, and clutches it frantically. His friends close in about him, and with the feeling that we have no right to gaze upon a scene that has become exquisitely painful, we hurry down the hill.

Somewhere, on these steep slopes, the mortal remains of our Lord's ancestors moldered back to clay ages ago, and His pious foster-father was laid to rest, perhaps by the loving hands of Him who wept at the grave of Lazarus, and in the death-agony committed His widowed mother to the care of His best-beloved disciple.

CHAPTER XIV.

FROM NAIN TO JEZREEL.

THE leaves of our Bible turn heavily, being damp with coming rain, as, guided by the "Biblical Gazetteer" at the back, we look up Luke vii. 11-15, and read that on the day succeeding the healing of the centurion's servant, Our Lord "went into a city called Nain, and many of his disciples went with him and much people."

"Now when he came nigh unto the gate of the city, behold there was a dead man carried out, the only son of his mother and she was a widow, and much people of the city was with her."

"Where is the city?" we interrupt the reading to inquire. David points to a cluster of wretched huts, walled with clay and roofed with straw plastered over with mud.

"A Moslem village!" we pronounce decidedly, as the scent of low-hanging smoke from hovel-fires is brought down by the moist wind.

Two kinds of fuel are in general use in such hamlets, chaff, and cakes of dried camel's manure, the latter predominating in regions where the ungainly "ship of the desert" is brought into the service of man. One of the most important branches of woman's labor in these circumstances is collecting the manure, molding it into flat cakes, and spreading it to dry upon stones, walls, and house-tops. I have seen roofs literally tiled with the malodorous combustible, and whole rods of stone-wall caked with it.

We smell chaff now, and prefer it.

"Did I tell you," remarks Alcides, reining back Massoud into a conversational walk, "what Rev. Mr. Jamal, the Nazareth pastor, said of chaff fires? In the village ovens they never go out, day or night, but smolder and smoke continually. He says that is the meaning of the 'unquenchable fire,' into which the chaff is to be cast—the fire that is not allowed to go out."

"Quite so, sir!" This from David, who is within ear-shot. "Matthew, third chapter and twelfth verse. Likewise Luke, the third chapter and seventeenth verse. They are the words of John the Baptist. The end of chaff in this country is to be burned. I do not know what the poorer people would do without it."

We are gazing at the knot of poor cabins that look as if a sweeping rain might wash them down into the plain some winter night, and marveling that no sign of a walled town remains, by the gate of which the funeral train may have emerged on the way to the rocky tombs we are told lie further down, and not so remote from the old Nazareth road as the ruins of the city—when something moves quickly across the lower plain directly within our field of vision. Four men hold aloft a misshapen burden, and half a dozen others keep pace with them in the long lope, swifter than a slow run would be. Serkeese grunts and says something we judge to be contemptuous.

"ALLAH IS GOOD!"

David translates:

"He says 'they have done up the dead man to look like our luncheon tent'—and the boy is right."

The luncheon-tent being swung across the very mule the "boy" bestrides at this moment, we can testify to the truth of the criticism. It is, nevertheless, a human body which the "fellaheen" (Syrian peasants) are carrying to the most desolate tract of a territory unrelieved by tree or shrub for many acres. Turning our horses in the direction they have taken, we come up with them as they deposit the figure, swathed in coarse matting, upon the naked earth, and begin with rude mattocks to make a hole among the stones—I cannot call the business "grave-digging." There are no mounds in what we now perceive is a cemetery, but ovals of round stones are set thickly all over the place, each designating a grave, and at long intervals we descry a low, upright block of granite, unlettered. The man who directs the gruesome task answers readily and civilly the few questions put through the interpreter.

"It is a child who has died. A boy of eight. He has escaped much trouble. He is better off here than living. Allah is good!"

Other figures appear from behind the shoulders of the bleak hills, women, bare-legged, but shrouded as to heads and faces in the dark-blue cotton mantles that uniform matrons and girls of the lower classes. They walk well and fleetly, and, approaching the grave-makers, without evincing the slightest curiosity at sight of us, seat themselves upon the ground about the motionless bundle and begin to sob in a subdued key, breaking finally into a chant, without beginning, middle, end, or the remotest suggestion of musical idea. The men pay no attention whatever to the performance. We characterize it silently as "the people making a noise," very much the same, doubtless, as was kept up by the "much people of the city," winding down from the Nain which is no more, when the Master's hand fell upon the bier and those that bare it stood still. One woman, as we note, neither sobs nor chants, but, her muffled face bowed upon her knees, sits at the head of the dead boy, motionless, as if as deaf as himself to all that passes.

David explains aside that the women of the neighborhood take advantage of each funeral to visit the graves of their own relatives, and shows us other and smaller groups, now arriving and scattered over the face of the hill. All crouch upon the earth soaked by yesterday's rain, and the humid air makes the dissonant croon of each family, "mourning by itself apart," audible to all the rest. Almost under Massoud's forefeet, one woman has thrown herself face downward beside a long, egg-shaped round of stones, and cries out, over and over, the same form of words to the newly-turned earth. David hearkens attentively, and interprets, line by line, in his grave undertone. Over against this translation, Alcides—having what he chooses to call, "a fatal facility for rhyme"—pencils a metrical version almost as literal:

I arose and came to the grave, at the breaking of day,
To the grave on the barren hillside where Abou, my husband, lay;
I called on him by his name, but no answer came again;
Once more I repeated the cry, but repeated it in vain.

No dear voice answered my cry, from the rough, forbidding stones.
He sleeps too deep beneath the earth to hark the widow's moans.
Who will help his children now, since my calls are all unheard?
O, form in the grave there, among the rocks! answer a single word!

The workmen step out of the shallow hole they have scooped, and lay the mattocks aside. Another relay covers the bottom of the pit with stones, and lays the uncoffined body upon them. Earth is thrown in until it is nearly level with the surface, and a second layer of larger stones tc hinder the depredations of jackals and hyenas; more earth, which is beaten hard, and the ring of great pebbles defines the last home of the little fellow who "has escaped much trouble."

We were surprised, but not shocked, to learn that our informant—the dignified man in a brown-and-white "abieh" and gray beard—is the father of the dead boy. While the grave is filling, an aged man, whom we take to be a priest, sits near by upon a gravestone and mutters fast some form of prayer, his palms outward. Four old men respond, when he pauses for breath, their hands outspread in like manner.

Now that all is over, the priest rises, and approaching the father, kisses him solemnly upon both cheeks, an example imitated by all the men present. The mother has not stirred, nor has any one spoken to her that we have been able to see, when we ride away and resume our journey.

A FELLAHEEN SOWER.

The funeral must have drawn to a centre all the population of the surrounding country, for during the next two hours we pass no human being except a swarthy well-built "fellaheen" who is sowing grain in a field lately scratched over by the wooden plough used by the so-named husbandmen of Palestine. He has made a "lap" of his outer garment, and it is full of grain. There is promise of an abundance of rain to settle the furrows, and he may be that phenomenon in this old land—a "forehanded" man.

Tabor lifts a gently rounded crest upon our left, and the wide plain passed an hour ago was the field in which Barak and Deborah overcame Sisera. Whenever we rise out of the occasional "dips" in the road, which is nothing but a bridle-path—we look for and find the snowy brow of Mount Hermon upon the horizon.

"Which Hermon the Sidonians call Sirion, and the Amorites call it Shener," writes the author of Deuteronomy, parenthetically.

At Shunem we pause to get a picture that we do not, for a moment, attempt to believe is the house of Elisha, and to arrange to our satisfaction in which of the no longer fertile fields the boy, now grown to be a lad, was "making believe" help the reapers when the fierce sun struck the young head.

"Did you ever think"—one muses aloud—"what a risk the poor mother ran in venturing out in the terrible midday heat that had killed her child? And in excitement and haste that were in themselves dangerous?"

There is no immediate response, but we know that the thoughts of all are with the solitary figure crouched together at the head of the flat, nameless grave in sight of the hill on which once stood Nain, and that each ponders upon the mother-love which is one and the same to-day with that which cast the Shunammite woman, bitter of soul and speechless, at the prophet's feet. Our luncheon-tent is pitched within the ancient bounds of Naboth's vineyard. The threatening rain-clouds, after a spiteful shower, have rolled by to mass themselves beyond Hermon and the hill Mizar. The sun shines kindly upon a spot which no man in his senses would covet at this day. Earlier comers have preempted the ground about the well that may have been here and used to irrigate the vines when Ahab went down to survey the blood-bought plot, and met Elijah

SYRIAN BREAD AND CAKE VENDERS.

face to face. One of my pet abominations, a camel, an uncommonly ugly specimen at that, erects his head against the pale sky. His master, his master's wife and his master's child, and the family donkey are enjoying their "nooning." The human creatures stare stupidly at us; the donkey is stolidly indifferent to our vicinity; the camel is supercilious. Hardy grasses push their way between the stones of every form and size that cumber the soil, and rank weeds fringe a lazy pool where a woman sits upon her heels, alternately beating wet clothes against the rocks, and going through the form of rubbing them between her fists while she gazes over her shoulder at the "foreign animals" feeding in their tent.

The Plain of Esdraelon stretches beyond to the foot of the mountains; the gleam of standing water gives back the sunlight at a dozen different points; copses of scrub trees diversify the level growth. It is not a barren outlook; neither is it attractive.

"A garden of herbs!" reads Alcides from "the best Baedeker for Palestine." "That didn't mean parsley, summer-savory, thyme, sweet marjoram, or even rosemary-and-rue, I take it—hey, David?

"I have often reflected upon that, sir, and in my humble judgment, King Ahab had a fancy for ornamental flower-gardens—terraces and the like—that could have been well contrived here. He was a spoiled child—was King Ahab—and his wife kept him spoiled—and a child—for reasons of her own. What are we to think of a man of years, and the king of a great nation, who takes to his bed and will not eat his dinner because he cannot get a bit of land he has set his heart upon? There was something evil at work besides his own folly. And when the evil outside of a man is a woman——!"

He shakes the crumbs vigorously from the tablecloth at the tent door with the gesture of one aware of the impotency of speech.

The vineyard was close under the walls of Ahab's palace or "house," as he called it in bargaining with Naboth. There is as little vestige of one as the other now. From out of the shelter of the tent, as we lie luxuriously upon our rugs, we gaze up at all that is left of Jezreel—a ruined watch-tower. From that height the watchman spied the armed company approaching and recognized the foremost as Jehu, the son of Nimshi, "for he driveth furiously."

"The marginal reading has it 'in madness,'" interpolates Alcides. "Hark!"

The thunder of racing hoofs is in the air; across the base of the hill rushes a drove of half-wild horses, bred upon the Plain of Esdraelon; leaping tall boulders and broad pools; tearing up the slope within pistol range of us, wheeling, as at a spoken order, and dashing off, until lost in the scrub.

We laugh, consciously, before confessing that for an instant we had seen, as by a flash, the mad driving of the rude usurper, the intrepid old queen, her eyes defiant between the painted lids, the tiara upon her wrinkled forehead, looking from her high window to fling her last taunt, barbed with a threat, at the soldier of fortune in her court-yard.

Alcides kicks at a cur who is nosing about in the grass for scraps, and shudders.

"I don't fancy the breed—*here!*"

The text he has in his mind is upon the open page before our eyes:

"In the portion of Jezreel, shall dogs eat the flesh of Jezebel."

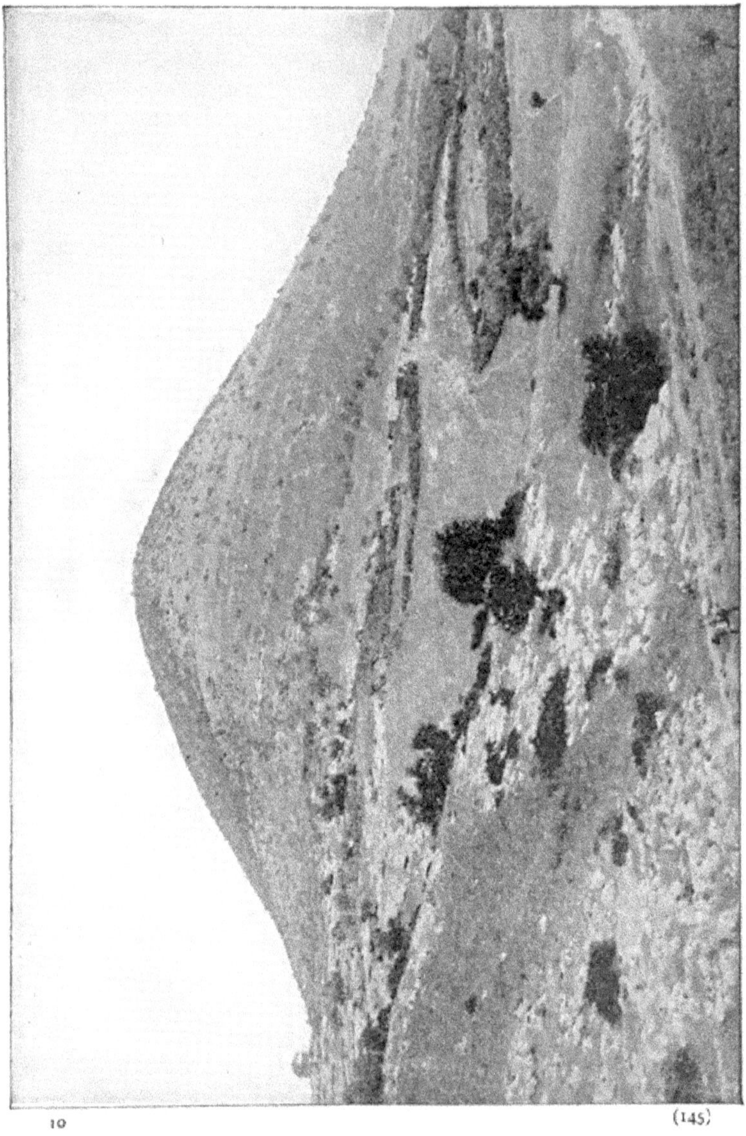

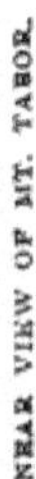

NEAR VIEW OF MT. TABOR.

CHAPTER XV.

GIDEON'S FOUNTAIN AND DOTHAN.

WE diverge from the regular route connecting Jezreel with Nablous, because Alcides has a strong desire, dating from his childish liking for "Gideon's band," to visit Gideon's Fountain. It is but a mile or so from our road. The rocks bound it on one side; on the other marsh lands stretch away into arid slopes which lose themselves in the barren bleakness of the mountains of Gilboa. Had David lived in this century, he would not have troubled himself to call down the curse of dewless and rainless seasons upon the naked sides and bald brows.

The "Fountain" is a shallow pool, perhaps sixty feet square, having its source in the rocks and finding an outlet in the sluggish brook. Of course, everybody alights for a drink at the spot where we imagine Gideon may have tested the qualifications of his host for the expedition of the morrow, and equally of course, nobody gets down upon his knees upon the brink to put his lips to the water. Then the horses are ridden slowly across the fountain and back again to wash and cool their hoofs.

"Dr. Sharpe does not believe that this is the pool of Gideon," remarks our guide, letting Dervish stand for a grateful minute up to the fetlocks in the gliding stream. "And Dr. Sharpe is a very learned man. However, we have always been told that it was here that the three hundred were chosen to march against the enemy encamped upon the Plain of Esdraelon. If the Captain"—a title newly bestowed by him upon Alcides—"will turn to Judges vi. 5, he will there read how they 'came up with their cattle and their tents, and they came as grasshoppers for multitude; for both they and their camels were without number; and they entered the land to destroy it.' No wonder Gideon was afraid to undertake the business of dispersing them. See verse fifteenth, if you please, sir—where he asks the Lord—'Wherewith shall I save Israel? Behold my family is poor in Manasseh, and I am the least in my father's house.' At that very time he was threshing his wheat when it was hardly ripe, to keep it out of the hands of the Midianites."

The clouds return after the rain, this time in a steady downpour, before we gain the shelter of the camp. It is pitched without the precinct of Jenin, a market-town for the surrounding country, situated near the southern point of the mighty oasis we have been traversing, and which, I may say here, becomes impassable during the January rains, to blossom with lily and rose and a riotous host of other wild flowers in March and April. Through the descending floods we made out the outlines of what looked like a barrack as we hurried into the tents. Before we have had time or means for drawing a dry breath we are advised that we are uncomfortably near an arm of law, if not of order, by the apparition of a couple of uniformed representatives. It is not easy to lay hand immediately upon the passports for which we have had no use since leaving Damascus, but it is presently apparent that they must be exhibited to the doughty officer who does

"LETTING DERVISH STAND UP TO HIS FETLOCKS IN THE WATER."

the talking. He may be the embodiment of an armed constabulary of his city, or responsible to his government for all suspicious strangers found loitering under the walls of Jenin. He is peremptory enough to be a triune mayor, high sheriff and pasha, as he drips over John's cook stove, and rains Arabic gutturals upon the conductor of the obnoxious foreigners.

It is the first moment of actual discomfort we have had in camp, for everything is wet that the rain could reach, and Syrian rain is a force to be remembered when once encountered. But a damp portmanteau is hastily overhauled, and from the very bottom is dragged a long envelope, marked "Passports," armed with which Alcides dashes across the rainy area to present it, defiantly, to his Highmightiness. David intercepts it, unfolds it in the kitchen-tent and extends it in dumb dignity. The official motions to Imbarak to hold a candle at his shoulder, and proceeds to peruse the document from top to bottom. Seen from the tent-door opposite, the scene has striking features, and Rembrandtish effects of light and gloom. We have time to study them all before the official snatches the candle and throws the gleam full upon the face of Alcides, standing unguardedly near.

"He has a beard!" dramatically.

If he does not add, "It is not so nominated in the bond," it is because Shakespeare is an unknown quantity in the list of his learning and accomplish-

"THE SPOT SELECTED BY OUR ATTENDANTS."

ments. He insists upon what amounts to the same thing, and so doggedly that, for the only time upon record, our worthy dragoman for a season bids farewell to his temper, and absolutely talks his countryman into shamed silence.

This is the first insult offered his travelers, and he means that it shall be the last, if he has to complain to the government in person of the stupidity of a man who pretends to serve it, and does not know that a gentleman has a right to shave his chin, or to let his beard grow, as pleases him, without asking permission of the Jenin police. No! the gentleman will *not* go up to the town, or one step out of this tent, to have his passport "viséd." The official knows, as well as the speaker, that the office where this should be done is locked up for the night. If he—the official—wants to pocket the contemptible "beschlik" (sixpence) it would cost to "visé" the "tezkere" (passport), the speaker will pay it to him out of hand, and here, to end a matter that disgraces the honorable government under which both parties live.

The conclusion of the whole matter is the ignominious retreat of the intruder, followed by his subordinate, who has not uttered a sound throughout the affair, and the business of camp-life can go on in the customary manner. David's spirit is too deeply ruffled to subside at once and as we sink to slumber three hours thereafter, we catch, between the dashes of rain upon the canvas roof, the murmur, not loud, but deep, of his rehearsal of the grievance to his lieutenants, as they smoke together in the kitchen-tent.

Camp is broken next morning, and the cavalcade sets forward under a sky as softly blue as ever bowed over Italy. Our first halt is at "Joseph's Well."

A GROUP OUTSIDE THE MILL.

"What we believe is the pit into which his brethren put him," is our introduction to the walled quadrangle and old stone building toward which the horses turn of their own accord.

"But we surely passed such a pit four days ago?"

"The Mohammedans are responsible for that blunder, sir. They claim that Joseph was let down into that well. Whereas, as every reader of the Old Testament knows, we learn in the thirty-seventh chapter of Genesis, the seventeenth verse, that Joseph's brethren were in Dothan. This is Dothan."

As to the last statement, there is no question. We exchange incredulous glances as, after having left the horses with Serkeese, severely enjoining him to keep his wits and all the bridles well in hand, we enter a gap in the low stone wall and perceive that the building it surrounds is a mill, erected above a square well. A water-wheel turns creakingly within this, constructed after a primitive pattern, jars being bound to the broad rim, going down empty and coming up full. The place is very dark until our eyes become accustomed to the change from the outer day; we walk up to the "pit," and look down into it.

"'And the pit was empty; there was no water in it!' Is it ever dry now?"

The miller holds up his hand in deprecation. What would then become of his business? And the visitors can see for themselves how many bring hither corn to be ground.

They are sitting about on stones outside, genuine children of Ishmael, albeit lounging against the wall precisely as Yankee and Western loungers haunt grist-mill or "store."

"STUDY OF REUBEN."—GEN. XXXVII. 29-30.

"The sight makes one quite homesick!" declares Alcides, pensively. "All they lack to complete the picture is the national jack-knife and shingle."

The one exception to what passes with the saturnine race for sociability is a solitary wrapped in his abieh and withdrawn to a corner of the mill-yard. Beside him on the wall are the trappings of a horse or mule. He regards us gloomily from the shadows of his kafeyeh. In the kodak-book he is registered as a "*Study of Reuben.*"—*Gen. xxxvii. 29-30.*

Try as we will, we cannot take Joseph's Pit seriously. There are too many of them. There is one in Egypt, the peculiar property of the Copts.

The place has other associations of which we speak softly to one another, after we have lunched, and David and Serkeese are breaking their fast in the shadow of the old walls. The country is fertile and beautiful with olive orchards. The wet trees glisten in the morning light, as from a bath of silver, and now, as far as the eye can reach, clothe the plain, and round off the heights. Hermon shows faint but fair, in the remote distance, already a belovéd landmark for which we look upon rising each hill. Peasants are plowing the brown fields with patient oxen, flocks of crows—gray instead of black—hopping and foraging in the shallow furrows.

In these meadows the sons of Jacob fed their flock, and compared notes concerning the little brother who was taking unwarrantable airs upon himself. We picture them herding morosely together at sight of the slender figure crossing the plain, appearing and reappearing under the low olive boughs. They would know him, when yet a long way off, by the many-colored coat. Did he fling it back as he ran, as the fellaheen whose salute we return as he goes by to the mill-yard, lets his brown-and-white abieh slip from one shoulder?

The boy had told tales out of school to their old father of the misdoings of his half-brothers, but his dreams were a more grievous offence in their sight. If—our mood more confiding with our talk—they did, indeed, cast him into the (then dry) pit behind us, they must have awaited his approach about where we were sitting upon the sun-warmed rocks, and in sight of the old caravan road from Gilead to Egypt. They were eating their luncheon (maybe nigh unto the spot selected by our attendants to-day,—who knows?) when they espied the dark line of Ishmaelites with their camels, bearing spicery and balm and myrrh, going to carry it to the richest land in the known world.

The Dothan—compassed at night by a host, both with horses and chariots, sent to take the prophet who told the King of Israel the words spoken by the King of Syria in his bed-chamber—has disappeared from the face of the earth with lesser and greater cities of the Holy Land. We do not know which of the picturesque hills bounding our view was the "mountain full of horses and chariots of fire round about Elisha," made visible to the eyes of the terrified servant in answer to his master's prayer. But the way by which the blinded Syrian troops were led by Elisha to Samaria must be nearly, if not quite, the route we are to take this afternoon.

CHAPTER XVI.

"THE BURDEN OF SAMARIA."

AND, I will make Samaria as an heap of the field, and as plantings of a vineyard; and I will pour down the stones thereof into the valley, and I will discover the foundations thereof."

We are climbing the hill that has stood from age to age, in ghastly fulfilment of a prophecy pronounced against the haughty rival of Jerusalem, two thousand six hundred years ago. The turns of the road are made by buried ruins; here and there a carved fragment has been washed bare, the corner of a stone sarcophagus projects. The thoroughfare is ancient, and probably the same by which the four leprous men stole down in the twilight from the gate of the besieged city into the camp of the Syrians, in the desperation of famine:

"If they save us alive, we shall live; and if they kill us, we shall but die."

The route is as solitary and silent now as when the night hid from the trembling outcasts the deserted tents and ways of what they had expected to find teeming with life. Of the old Samaria, nothing is left but heaps upon heaps, and, bordering the neglected road upon a plateau more than half-way up the mountain, a line of truncated granite columns, the remains, say some, of the market-place, or forum, of Omri's city.

"If you please, Captain, you will find that in First Kings, the sixteenth chapter, twenty-fourth verse," interjects a respectful voice from the head of the little cavalcade. The Bible cushion in these wayside lectures is the pommel of the saddle generously pressed upon Alcides by the belovéd physician in Beirût as infinitely more comfortable than the ordinary Arabian. From the peripatetic desk we hear the first record made of the celebrated town.

"And he bought the hill Samaria of Shemer for two talents of silver, and built on the hill, and called the name of the city, which he built, after the name of Shemer, owner of the hill, 'Samaria,'" or, as the margin has it, "Shomoron," or, "Watchfort."

An orchard of olive-trees backs the ruined pillars, and a little further on the road ends in the modern village of Samaria.

We have only time to see that it like a majority of other Syrian hamlets—less thrifty than Nazareth, and not so large and cleaner than Jenin, when our guide rides back from the front to the spot where we have paused for a view of the landscape, and the shining ribbon of the Mediterranean bounding it upon our right.

"If you please, here is something you will like to see,—the boys of Samaria at the very games, I doubt not, which the holy men of old, when children, played in Israel and Samaria, and in like manner."

We hasten into an open space,—the village common, it may be called—peopled, at this instant, with a wild rabble of youngsters, scratching, biting, kicking and

"CARRYING THE MONEY WITH THEM."

swearing (there is no mistaking the tone of profanity, however strange the tongue), like a pack of quarrelsome hounds.

Jeremiah would have laughed with us, at the unexpected climax and the added stimulus of the grieved chagrin eloquent in the honest bronzed visage of the showman at our shout of "Holy men of old! Oh, David!"

"And they were so peaceful and friendly not three minutes ago, I do assure you!"

They are neither friendly nor reasonable now, for when they can be persuaded to stop fighting one another, they despise the admonition to "go on playing," and after basely accepting a bribe, get into position by driving a stick into the ground, then, catching sight of the kodak, raise a shriek of "sorcerer!" and pelt up the hill, carrying the money with them.

Our respect for the Crusaders deepens with each visit to such memorials of their occupancy of the Sacred Soil as the ruins of magnificent bridges, strong fortresses and noble churches which the wanton Saracen and pitiless Time have not succeeded in effacing utterly. The impression is too general, even among educated Christians, that the impetuous zeal to rescue the Holy Sepulchre from the hands of the infidel expended itself in a great measure in spasmodic fury such as was displayed by the bereaved bridegroom who—

Threw his life away, fighting with the Turk.

Treasure and taste were lavished, no less than the best blood of Christendom, in the effort to establish the true faith in the land that had given the Founder of Faith to the whole world, and it is not in human nature not to grow bitter, sometimes, at thought and sight of the desecration that has succeeded a sublime, if ill-judged, enterprise.

A fondly-cherished family tradition has to do with a renowned champion of the Cross, who fought every inch of his way from Jaffa to Jerusalem, and left his bones in the dust of Mount Zion. We have never acknowledged the kinship more fully than in the indignant leap of our pulses at an incident of our visit to the Church of St. John the Baptist, the one object of interest in modern Samaria. The walls have withstood the action of almost nine centuries, "Crusader's work" being the synonym for stability from corner to cap-stone. Hammer and trowel are busy upon the interior as we cross the marble threshold. The whole church is gutted, and is undergoing alteration into a Mohammedan mosque. Portions of it have been thus used for a hundred years. Under the hands now engaged here, the last vestige of Christian architecture will, in a few years, be obliterated. Holding, as we do, as a rule of good breeding, the obligation to bear ourselves decorously in whatever place of worship we may visit, we have made the round of the place quietly, and are on our way to the entrance, when a man who seems to be the superintendent of the workmen steps up in front of us and addresses us in Arabic, to which David replies.

We are evidently the objects of remark in the ensuing dialogue in which the stranger presses some point, and Jamal as resolutely, although more temperately, refuses.

"What does he wish to have you tell us?" we demand, having made out this much of the tenor of the altercation.

"Nothing worth my telling—or you knowing," is the answer. "It is better I should not say it. He has no right to speak it to me, much less to ask me to say it in your ears."

The excitement of the Moslem rises dangerously fast, and we, too, make a stand. "What have we done?"

"Nothing whatever," the interpreter hastens to convince us. "If you will pardon me for carrying such words to your ears, he will have it that I tell you that the faithful (as he will call them) have, at last, broken down and cast down the signs of idolatrous religion that were once in this church, and that the like should be done everywhere, as he hopes and he believes it will be before long.

"THE DREARY LINE OF NAMELESS COLUMNS."

That is all; and now, if you please, we will thank the gentleman for his hospitality and leave him."

We take this advice, without trespassing further upon Moslem courtesy by asking to see the reputed tomb of John the Baptist in the crypt.

"It is by no means probable that he was buried here," we are informed. "It is believed that John was beheaded at Herod's castle at Machias beyond the Jordan, and as the Captain will see in St. Matthew xiv. 12, and likewise in St. Mark vi. 29, his disciples got possession of his body and buried it. St. Mark says, 'laid it in the tomb.' Then 'they went and told Jesus.' They would hardly bring him to Samaria. There was a splendid temple here then, built by Herod, I have heard, and dedicated to the Emperor Augustus Cæsar."

He pauses to speak to a woman sitting upon the ground, an earthenware bowl of dough between her knees, in the door of a stone building plastered with mud. She replies to his kind tone, testily, without looking at him, and draws her veil over the lower part of her face. We know the place for a public oven, having seen such elsewhere, and having, also, witnessed the friendly favor gained by our dragoman with all sorts and conditions of people, are amazed that this acquaintance should snub him decidedly. He shakes his head sorrowfully in walking away.

"That woman's husband was sick last year, and I was kind to him, and to-day it is she who has told the boys that we are sorcerers, on account of the kodak, and she says she is herself afraid of it. Sorcery! and in the nineteenth century."

Apparently, it is the Christians with whom the Samaritans, now-a-days, will have no dealings.

Beyond the dreary line of nameless columns lining the ancient market-place and the olive orchard, amid a thicket of low trees—known by Christian peasants as the "Christ-thorn," because of the legend that from a branch of this the thorny crown was plaited—and a tangle of furzy growth, gathered when dry for fuel by the poorest women,—is the alleged site of the "ivory palace of Ahab's queen."

Almost a century after the memory of the wicked sovereign and his wickeder consort had rotted from the mind of the nation they had misruled, the herdsman of Tekoa saw in a vision the fairy-like structure that was the coronet of mighty Samaria, and prophesied concerning it:

"And I will smite the winter house with the summer house; and the houses of ivory shall perish, and the great houses shall have an end, saith the Lord."

Lizards, green and black, slip in and out of the jungle, scampering faster as the branches are stirred by a lean, starved-looking donkey, who pokes his head between two thorn-bushes to look at us. We have seen no drearier spot, unless it were the heap of stones passed soon after leaving Damascus, a cairn covering the reputed grave of Nimrod.

No one who has read the weird lines could refrain from reciting aloud here:

> They say the Lion and Lizard keep
> The courts where Jamshyd gloried and drank deep;
> And Bahram, that great Hunter, the wild Ass,
> Stamps o'er his head, but cannot break his sleep.

The prickly furze—for which we know no other name—is at the driest now, and Bedouin women, from the encampment of booths we see in a hollow of the great hill down there, are cutting it with bill-hooks, binding it into immense bundles and bearing it home upon their heads. It is light, although so bulky; yet we watch them with wonder, stepping fleetly over stones and stubble, straight and strong, apparently careless of the loads.

In the direction toward which our faces are turned a panorama of fertile loveliness unfolds itself to the delight of eyes wearied by sterile terraces and jutting

"THEY FALL INTO LINE."

crags. Slopes are greening under a day of blandest sunshine; in sheltered crevices, flowers—pink, white and yellow—are beginning to bloom, and in the gentle descent into the Valley of Shechem, the silver-green groves are vocal with the laughter and prattle of children who are gathering purple olives from laden boughs. They fall into line and return cur salute as we pass.

"The labor of the olive has not ceased," says one.

Yet, in the mournful recollection of the matted jungle overgrowing the fallen stones of the royal palace, another wondrous prophecy is read aloud before the "guide-book" is returned to the saddle-pocket:

"And it shall come to pass in that day that every place shall be, where there were a thousand vines and a thousand silverlings, it shall be even for briars and thorns;

"With arrows and with bows shall men come hither; because all the land shall become briars and thorns."

This is Isaiah's cry against "the head of Ephraim, which is Samaria."

And "For all this, His anger is not turned away, but His hand is stretched out still."

JACOB'S WELL BEFORE THE RECENT EXCAVATIONS, EBAL AND GERIZIM IN BACKGROUND.

CHAPTER XVII.

JACOB'S WELL AND JOSEPH'S TOMB.

PURSUING our way from Nablous (the ancient Shechem) in the cold winter morning, we are spared the sight of the lepers who sit daily at the wayside, begging, when the weather is moderate. Cold is peculiarly painful to these unfortunates, and until it abates, they lie as nearly torpid as their condition will allow, in whatever shelter they can find. We see but four to-day. In the immediate outskirts of the town, one, better dressed than a majority of his class, his face muffled from the sharp air so that but one eye is visible, and with his maimed hands behind him, shivers against a sunny wall, and another is huddled in a neighboring doorway. When we leave the streets behind us we pass two more; one squatted on the stones, his companion lying upon his stomach, supported by his elbows.

All stare vacantly at us, too demoralized by the fall of temperature to unclose their lips, even to utter "bakshish." If any men and women upon this beautiful earth have a valid excuse for beggary, it is the lepers of the country towns. Cut off from such branches of industry as require people to labor in companies, debarred by mutilated members and stiffened joints from tilling the fields, their only resort from starvation is to cry aloud and spare not to entreat the charity of all who pass by. If the day were mild, both sides of the road would be lined with them, and the strange, husky call we learned, long ago, to recognize above the clamor of crowded thoroughfares—"*Bakshish! howadji! Abras! Abras! bakshish!*" (Give money, gentlemen! we are lepers! lepers! Oh! give money!")—would deafen the ear and sicken the heart. Whatever may be our views on the subject of mendicancy and the evils of indiscriminate alms-giving, we cannot refrain from throwing some coins to them in hurrying by, and looking back, see them stoop slowly and painfully to pick them up.

This is the sharpest morning we have felt. John, of the numerous appellations, and Imbarak, the tall, following the mules laden with tents and camp-furniture, are bundled up in all manner of extemporized shawls and scarfs; the jaunty waiter, erect as a ramrod in the discharge of official duties, is bowed and shrunken as a gladiolus stalk under a nipping frost, and glances piteously at us in response to our cheery "good-bye."

The weather that shrivels up these children of the sun has cleared the atmosphere from horizon to horizon. The valley of Shechem is wondrously beautiful in the pure light. But we miss Hermon, gleaming as with newly-fallen snows, and the purple peak of Safed, the highest point of Galilee, so often visible to us for the past two days.

"To which, it is said, Christ our Lord pointed when he said, 'A city set on a hill cannot be hid!'" is our guide's comment upon the storied height.

Every foot of ground is connected with the sacred Classics. For while Nablous means nothing to us, our eyes, ears and hearts are opened when we recollect that the modern town is the ancient Shechem. Mount Ebal, on our left, and Gerizim, on the right, are the Mounts of Cursing and of Blessing.

"WE PASS TWO MORE."

"If the Captain will kindly turn to Joshua, the eighth chapter, thirty-third and thirty-fourth verses," suggests our animate Concordance. In the still, dry air, the words have a peculiar ring that helps us to believe what learned commentators upon the remarkable narrative aver, namely, that the responses chanted by that half of the Israelites stationed upon Mount Ebal could be heard across the narrow valley upon the summit of Mount Gerizim:

And all Israel and their elders, and officers, and their judges, stood on this side the ark and on that side before the priests the Levites, which bare the ark of the covenant of the Lord, as well the stranger as he that was born among them: half of them over against Mount Gerizim, and half of them over against Mount Ebal, as Moses the servant of the Lord had commanded before, that they should bless the people of Israel.

And afterward he read all the words of the law, the blessings and cursings according to all that is written in the book of the law.

"What a magnificent responsive service!" observes one, as the "best Baedeker for Palestine;" is closed.

Not put away, for Shechem of old figures often and prominently upon the inspired pages; the mention of the name draws about it a host of striking pictures. In Mrs. Whitney's one book of foreign travels, "Sights and Insights," she puts into the mouth of Emory Ann, a shrewd New England spinster, a saying we borrow often, and at last appropriate unscrupulously. She describes herself as "realizing her geography." We are "realizing our Bible."

"JOHN AND IMBARAK."

The little village whitening in the sunshine and nestling at the base of Gerizim marks the place where the Father of the Faithful built the first altar raised to the One True and Only God in the land then held by the Canaanites. They call it now, "The Holy Oak," in memory of the tree under which Jacob buried the "strange gods that were in their hands, and all their earrings that were in their ears," and where Joshua set up the great memorial stone of the "statute and ordinance" made in Shechem.

Nearer to us, and toward this holy site, we turn our steps in mute reverence. For this is "the parcel of ground that Jacob gave to his son Joseph, the parcel of a field where he had spread his tent, which he bought of the children of Hamor, Shechem's father, for an hundred pieces of money." We turn from Genesis to John's Gospel.

> Now Jacob's well was there. Jesus, therefore, being wearied with His journey, sat thus on the well; and it was about the sixth hour.

The "sixth hour" would be high noon, and there had been a fatiguing walk over the hills. As He "sat thus" (wearily), upon the curb of the well, He

"DAVID RELATES, SEATED ON THE UPPER STEP."

needed the meat which his disciples had gone into the city to buy, no less than water and rest.

A rough, broken stone wall surrounds the well. The archway is closed half the way up by a door. This is locked, and there is no care-taker near to admit us. Selecting the lowest breach in the wall, we clamber over fallen masonry to the top and over it to the inside. A church stood here once, probably erected by those valiant and indefatigable church-builders, the Crusaders; but altar, columns and architrave were swept away centuries ago, and the mosaic pavement buried many feet deep under the rubbish we see. A feeble effort at exploration for what we know must underlie the ruins was made recently, and laid bare the steps. A short flight leads down to the landing; beyond the landing is another locked door. We shake it energetically and uselessly. The priest who acts as custodian is absent, and the key is with him. On the other side of the door is the mouth of the well, more than seventy feet deep.

"It was, we are told, over an hundred feet in depth in former times," David relates, seated on the upper step. "The custom of visitors of throwing stones to

JOSEPH'S TOMB.

the bottom to hear them fall into the water has helped fill it up. There used to be a flat stone on top, with a round hole in the middle, through which pig-skin buckets were let down by long ropes. I'm thinking that a few thousand dollars would be well spent in clearing away all this rubbish, and show what the well was on that day."

The sigh is continually in the mouth of the visitor who bears in memory that the deposit of ages has arisen many feet above the surface upon which our Master walked. We make our way up to the present level; lay our Guide-book upon the wall, and re-read the beautiful story of the interview with the woman of Samaria.

"It is called Askar, now," we learn, "and lies just there"—a finger is pointed toward Mount Ebal—"a little distance away."

It was an insignificant town then, and she who came carelessly along the path winding around the hill was not a good woman, an ornament to the place. So loose had been her life, and so bold must have been the glance she dropped upon the Galilean peasant, sitting with lax limbs and drooping head—maybe upon the massive steps we have just left—that the impulse of a man who consulted self-respect rather than the opportunity to lift the fallen, would have been to feign ignorance of her presence.

"THE GIRL WE MET AT BANIAS."

We have never appreciated until at this reading the effrontery of her rejoinder to the request common to the lips of every way-worn man, halting at a well, from the days of Eleazar of Damascus down to this, our day—"Give me to drink." She might be notoriously profligate, getting rid of, or being cast off, by one after another of five husbands, and at length braving public opinion and risking a shameful death by living with one who was not her husband; but she had not sunk so low as to give a draught of cold water to a Jew. She was a Samaritan, through and through. Even when her insolence was quelled by the majesty of the Stranger's presence and speech, she vaunted herself upon her great ancestor, Jacob, and the patriarch fathers who worshiped in "this mountain."

"That would be Gerizim," we check the narrative to say, looking up with awe to the summit on which may be traced the ruins of an ancient synagogue. However changed other features of the scene may be, the hills are the same, and it is certain upon what His eyes rested as He spoke of "this mountain."

We have sent the horses on to wait for us at the Tomb of Joseph, and walk the quarter-mile separating us from Jacob's Well. The low-domed mosque beside the last resting-place of the patriarch glares whitely against the back ground of mountains. The tomb itself is in the enclosure without the mosque,

GATE OF BANIAS (THE ANCIENT DAN).

and olive trees grow so near as to shade it when the sunbeams slant from the west.

"For over sixty years his mummy must have traveled with the tribes;" Alcides has been making a rough computation upon the fly-leaf of his note-book, based upon the chronological headings of certain pages in the "Palestine Baedeker." "And he had lain in a coffin in Egypt four hundred years. Not such a long time compared with the two or three-thousand-year-old mummies one makes free with in museums; yet one cannot but feel that it must have been a comfort when Joshua at last buried his bones quietly (and for all time, let us hope) in this parcel of ground."

"A JOSEPH OF TO-DAY."

It is a goodly resting-place, guarded by the everlasting hills, upon one of which Eleazar and Phineas, the son and grandson of Aaron, were interred.

The defile between the Mounts of Blessing and of Cursing, scarcely more than half a mile wide in the narrowest part, widens into what is styled the "Wandering Field of Joseph." Again—(and I am ashamed to confess for what number of times we might set it down)—we discover that our preconceived idea of time and distance in a familiar Bible story is utterly at fault. If catechised we should have said that Joseph's search lasted throughout, perhaps, one day, and that the ground covered by the "wandering in the field" was measured by acres, not miles. It absolutely startles us to discover that the seventeen-year-old lad had traveled over forty miles—measured by an air-line—after leaving his father at Hebron, before he met "a certain man" at Shechem, who had heard Jacob's elder sons say, "Let us go to Dothan." Dothan, itself, is fifty miles from Jacob's home in Hebron.

One is moved to compassion and admiration at the thought of the anxious hours and tedious days consumed in the errand, and the steadfastness that held the messenger to it.

"Yet," we remind one another, "the boy was used to such expeditions. Do you recollect the Bedouin girl we met at Banias, who had been looking for her father's camels for two days, and had not found them when we saw her? She had slept at night in the tents of her people, as she chanced to come up with them. Joseph no doubt did the same, or laid him down to rest, as darkness fell upon the grass. That his brethren were in the fields with their flocks shows that the season was not winter."

A shepherd-boy appears in the defile terminating in the "Wandering Field," swinging himself along with the springing tread of the mountain-bred native. His tunic is dark blue cloth, the upper garment of sheep-skin made up with the wool inside; a scarf of many colors is twisted about his waist, and a gay "kaeyeh" is bound over his head with a black "ikal," or cord of camel's hair. In the momentary pause caused by the sudden sight of the approaching cavalcade, the kodak has caught him as a "Joseph of to-day."

FOUNTAIN IN CANA OF GALILEE.

CHAPTER XVIII.

THE SONS OF ISHMAEL.

THE black tents clustered, in winter, in sheltered hollows of the hills have been, from the date of our entrance into the Holy Land, objects of peculiar interest to us. Like those of Kedar, they are comely in our observant eyes. Once and again, we have alighted at a small encampment, and sat for an hour in the tent of the sheikh, accepted coffee, and declined pipes and cigarettes upon one or another civil pretext, and talked (at second-hand) with the inmates of the temporary home. More frequently we have ridden near enough to witness the hospitable stir created by our approach, exchanged friendly salutations and passed on, for a twenty-minute or half-hour call is an impossibility without giving deep offence to the astonished host.

We are now to pay our first regular visit to a Bedouin sheikh, whose village lies within three miles of the table-land where we pitched our camp last night, and directly in line with our projected route for to-day.

The potentate, in passing our door at sunset, had alighted to proffer in person a courteous request that we would honor his poor abode by a call at whatever hour it might please us to do so. All that he has, and whatever he can do, is, according to his showing, at our service, now and forever. His grave dignity and graceful deference of tone precluded ungrateful doubts of his sincerity.

Without emulating his deportment, we accepted the invitation, leaving our major-domo, as master of ceremonies, to clothe the acquiescence in terms sufficiently high-flown to satisfy himself and our would-be entertainer. Jamal came to the dining-tent when dinner was over, to post us up as to the sheikh's rank and importance. He is one of the richest and most powerful Bedouins on this side of the Jordan, with a following of over five hundred families. His flocks of sheep, his herds of goats and larger cattle are immense, and his horses are the finest for many miles around.

"He has but two wives," is a part of the story, "He is only forty years of age, and may, of course, add to the number. Yet that is not likely. He is a sensible man."

Having witnessed the parting between the proprietor of our camp and the mighty sheikh, we quite comprehended the compliment of the call, and the friendliness of the pair.

We talk Bedouin until our early bed-time, comparing impressions and information with regard to the strange nomads. The two hours' discussion is thus boiled down in Alcides' note-book:

"Their object in living seems to be to rob other tribes, and to fight the injured parties afterward; to hunt, to ride, marry, and to wander from one chosen place to another. Their ostensible business is agriculture and stock-raising.

"Leading characteristics: Politeness and hospitality to guests; revenge and ill-doing to enemies, and a large and level eye to the main chance, especially in the matter of robbery and horse-trades.

"HAVING WITNESSED THE PARTING."

"In morals of other kinds, their methods are summary. If a man suspects his wife of undue liking for another man, he invites her to accompany him upon a hunting expedition, or a pleasure ride, and comes back without her. She is never seen or named again, and the murderer is honored for the deed. If a girl gets 'talked about,' there is no investigation of proofs. Her father or brother takes her off out of sight of the camp, and shoots her as he would a dog suspected of hydrophobia. Half-way measures are unpopular among this simple-minded people."

At breakfast, the theme is still the sons of Ishmael and the visit we are to make. Imbarak, dwarfing the tent, as usual, waits without, tea-pot in hand, five minutes before his presence is perceived. He is too decorous to "hem," or shuffle, to call attention to the unimportant circumstance of his existence. And this morning, alas! he casts no shadow. He never casts much, as to width, but the sun which arose red and sullen has, as the old saying is, "gone to bed again," under an uncompromising gray curtain.

The curtain lets down more closely, on our side of the nearer hills, by the time we are on the march. As we reach the one irregular street defined by the

TEA AT OUR OWN TENT DOOR.

low, black tents, preliminary drops have lengthened into business-like streams. It rains, and rains straight.

"They were meaning to remove the encampment to-morrow, up to the hills," David says. "As you will see, the rains have made this flat, low situation quite muddy."

There are perhaps one hundred tents, all low, and some long, covered with a coarse fabric, woven of mingled goat's and camel's hair. Oddly enough, although smoke is rising everywhere through this stuff as through mosquito-netting, it is practically water-proof. On the rainward side curtains are fastened to the ground. As the day is not cold, only wet, the other sides and ends of the tents are open. The settlement is sparse in the suburbs, and diversified by trees and bushes. As we go on, the tents are nearer together, almost touching one another. They vary from ten to forty feet in length, and from ten to fifteen in depth; the sides are five feet high at the lowest, and about eight from the earth to the ridge-pole running from end to end. The largest is, as a matter of course, the sheikh's, and is further designated by a spear stuck in the ground.

Before we halt, half-a-dozen men rush out, ready to seize the bridles of the horses and assist us in alighting. As each takes a visitor's hand to help him down, he tries to kiss it, with a gesture of profound humility. Alcides, whom David coaches upon nice points of Syrian etiquette, dexterously evades this ceremony by withdrawing his hand and touching the back of it to his own lips. A woman and a foreigner permits the salute, bowing and smiling appreciation of the compliment paid to her. We are escorted into the largest compartment of the tent, the reception, or guest-room. There are three divisions, that in which we are received, and one apiece for each wife and her children. In the middle of the earthen floor smoulders a fire of coals, and in the hot ashes are set three brass coffee-pots, the tops attached to them by slender chains.

The sheikh rises from a low stool at the left of the fire, and advances to welcome us, raising his hand first to his forehead, then to his heart. Behind him is a row of men, cross-legged upon rush matting unrolled from one end of the tent to another. All bow and touch their foreheads without rising. At the right of the fire, servants spread a crimson rug, thick and soft, then place in the centre of it a camel's saddle and housings, richly embroidered. Upon this, as the seat of honor, we are invited to place ourselves. Little is said while all this goes on, gestures, much salaaming and a few sedate Arabic gutturals doing the work of making us at home and comfortable. When we are seated, the attendants back to the rear wall of the tent, and stand, alert, but motionless, behind the line of sitting figures. These, we are now aware, are fellow-guests with ourselves, invited to do us honor. Almost simultaneously, we find out that we have been invited to breakfast. There are as many breakfasts per day in a Bedouin's lodge as there are relays of guests in the forenoon. If these arrive hourly, each hour a fresh breakfast is cooked and served. One of the true tales to which we hearkened overnight was of a Bedouin, who, pursued by those who sought his life, fled to the chief of the hostile tribe for refuge. The sheikh received and sheltered him, and asked what the fugitive would like to have for dinner.

"That," said the other, "I am content to leave to your generosity."

The sheikh left the tent for a few minutes, returning presently, and resuming his seat in tranquil gravity. After a while, the guest, growing uneasy at the bleating and bellowing going on without, sallied forth to see eighteen sheep and half-a-dozen calves lying bleeding on the grass, and the work of slaughter still going on. But for his interposition and vehement protests, all the live stock in the encampment would have been sacrificed upon the altar of hospitality.

The bill-of-fare on the present occasion was settled before our arrival. The table is set by depositing upon the floor a huge bowl of stout earthenware, filled with a smoking mess of broth, thickened with lentils until it would almost stand alone. About the basin is arranged upon the matting a circle of round, flat cakes of unleavened bread, hot from the oven or griddle on which they were baked. The central dish is flanked by a smaller of honey and one of loppered goat's milk, or lebben, the same delicacy offered by Jael to Sisera in a "lordly dish,"

"THE SETTLEMENT IS SPARSE IN THE SUBURBS."

probably no better as to quality than the brown ware in which the national article of diet is placed before us. As many guests as can conveniently gather about the "dishes" are collected by the host. They double their legs under them with the ease acquired by long practice, and each tearing off a fragment of the tough, warm cake, folds it with thumb and forefinger, plunges it into the reeking mess, and carries his spoil to his mouth. Portions of lebben and honey are partaken of between times.

Our guide, always considerate of our well-being, assures us, in a guarded undertone, that we need not eat unless we desire it. He will represent the late and hearty breakfast we have had, and make all right. Alcides politely waives the apology, takes a distinguished place in the ring and tears off one "spoon" after another from his round of bread, as to the manor born. It is excruciatingly funny to behold when one's stomach has nothing at stake in the exhibition, and the trifling accident of a brown thumb withdrawn, dripping, from the trencher, and carried with the bread-ladle to bearded lips, signifies less than if the next dip and sop were to feed the beholder's inner man—or woman.

Coffee-making goes on all the time. The raw berries are roasted upon an iron plate shaken by a long handle over the enlivened coals; then they are turned into a mortar made of oak from Bashan and of a rich brown with age and use, and pounded with a pestle by a servant. As he pounds and grinds, he beats a sort of tune upon the resonant sides of the mortar, monotonous and not unmusical. Lastly, the coffee powder is put into one of the brass kettles, simmering on

"TURNS HER TOWARD THE KODAK."

the fire, and boiled up three times, before it is poured into small cups, without handles, and passed to the guests. I taste it, and even drink half of the contents of the cup although it is thick, unsweetened, and made absolutely odious to the uninitiated by a liberal pinch of allspice. The exhilarating draught is dispensed five times during the two hours of our stay within the hospitable precincts. Cooking is carried on as industriously in the adjoining compartment, to a subdued accompaniment of women's and children's voices. In the presence of so many men, the wives and women-servants are not allowed to show themselves. In view of the prejudice—or principle—enjoining the seclusion, we cannot admire too heartily the perfection of breeding that extends to the pale-faced stranger of the inferior sex attention as marked as that received by her escorts. Not a glance, much less a word, testifies to surprise at her uncovered face and assumption of equality with the men surrounding her. Furthermore, not a question is asked as to the antecedents or intentions or destination of the foreigners. The talk runs (or walks) on the weather, the crops and horses, and is led by the sheikh, who eats apart from the rest, as a token of humility. He is not worthy, he intimates by the separation, to dip his fingers in the dish prepared for those he delights to honor.

When the first guests have satisfied their hunger, another set takes their place; the central bowl is carried away to be refilled, more bread is brought in, and the bowls of honey and lebben, if emptied, are replenished. Napkins, finger-bowls, spoons, knives, and forks are unheard of accessories of civilization. All the furniture in the "drawing-room" has been mentioned, and there is little more in the "hareem," or women's tent, except rugs and quilts used for bedding, and the few brass and earthenware vessels needed for primitive cookery.

"The Bedouin's life is not a happy one (in winter)," is scribbled upon a page of "Notes by the Wayside." The valleys and low-lying plains, selected because the winds there are less violent than upon higher ground, are at this season sodden with rain, and when the storm is violent, fowls and cattle rush pell-mell to the one available shelter, and will not be excluded. The floor is drenched by dripping fleeces and hides, and trodden into a quagmire by many hoofs, and it is not a rare circumstance for the human family to be driven to take refuge in one compartment, leaving the others for the dumb creatures. However crowded the premises, the guest, be he friend, stranger, or even foe, has the best place in the tent and the choicest portion of food, and the host would protect him at the risk of his life against insult or attack, were the assailant of his own tribe and kindred.

The parting token of hospitality that effaces self in the desire to gratify the visitor is given by our host, when, the sun breaking forth for a brief half-hour, we are in the act of leaving the encampment, and Alcides quietly tries to get a picture of a woman crossing the open space about the principal tent. The intention is perceived by the sheikh, who overtakes her with a few strides, lays a hand upon her shoulder and turns her toward the kodak. He probably has no suspicion that he also appears upon the plate; the attitude is one which his compeers would consider derogatory to the dignity of his sex and office. Nevertheless, we believe that his action would have been the same had he had time to weigh the consequences of the attempt to meet the wishes of the parting guest.

CHAPTER XIX.

THE SONS OF ISHMAEL.—(CONCLUDED.)

A LESS formal call than that reported in the last chapter is paid to a Bedouin camp of humbler pretensions and many miles distant from the powerful sheik whose portrait accompanied that chapter.

A woman came to our tents this morning before breakfast with exactly two fresh eggs to sell. She belonged to the poorer class of Bedouins, emigrants from the wealthier tribes across the Jordan. A thin, miserable-looking baby was upon her shoulder, and scratched its head while the kodak, hidden by a curtain of the tent door, took a picture of mother and child. She had come from an encampment four hills away, she said, and was on her way to Jericho to spend the money she would get for her fresh eggs. They were laid yesterday. John had his orders to pay her three times what they were worth, and to give something to the baby. With all my love for the young of the human species, I did not care to go nearer this specimen.

It is after our lunch and siesta, that our journeying brings us in sight of the village the woman must have quitted at dawn for the walk over the hills. The turn that reveals it, burrowing, as it were, between two sand-mountains, is so abrupt that the community is spellbound for an instant. Children playing upon the waste lands about the black booths, gape at the approaching party, and the one man visible out-of-doors is the only person who has the self-possession to call back the pack of yelping, baying curs rushing about the feet of the horses. To the momentary consternation succeeds a hurry-skurry on the part of the women, led on by a tall figure who is named to us as the principal wife of the sheik. From two compartments of the three-roomed tent they issue in frantic haste, laden with quilts and rugs, and bearing them into the guest-room. By the time we enter this, a sort of bed or divan is laid around two sides of the compartment, and cushions are piled at irregular intervals upon it. The wife-in-chief meets me, stoops to kiss my hand, carries it to her forehead and her heart, and indicates where she would have me sit. When I comply, but do not recline, another older woman pats the nearest heap of cushions, and smiles persuasively. A compromise is effected by resting an elbow upon this "arm" of the divan.

The sheik's favorite, registered in our note-book, at one time, as "Zenobia," at another, "Boadicea," must have been as nearly handsome in youth as a dusky maiden can be, whose forehead, cheeks and chin are tattoed with polka-dots, rings, crescents, stars and other designs, and whose upper and lower eyelids are painted, She is stately in shape and carriage, and wears her dark-blue gown and veil, upon which are some tatters and embroidery, as an empress her ermine. Taking a handful of dried thorns and twigs from a maid, she kneels over the hillock of white ashes in the middle of the floor, and herself kindles the fire, pushing up her loose sleeves to the elbow, and holding back her veil that she may blow the embers to a blaze. Her slender wrists are loaded with silver bracelets, and her arms are tattoed as far as we can see them. As she sits on her heels, watching the result of her labor upon the creeping flame, we open the conversation by thanking her for taking so much trouble to make us comfortable, and protesting against further exertion on her part. She answers without withdrawing her eyes from the fire or turning her head.

"MOTHER AND CHILD."

"She says that she is only performing her duty in her husband's absence," says David from the other side of the apartment. "That, should he be told, when he returns, that she had done less, he would be displeased with her. Also, that he would have a right to be."

Uneasy at the silence that prevails, we give the ball of talk another shove.

"Tell her how much we admire those bracelets, and that we have seen no others of that pattern. And if it is quite the thing, ask where we can find some like them."

"In Jerusalem!" returns the sheik-ess, laconically, still staring the fire out of countenance.

Seeing it, at length, burn brightly, she rises like a queen, for all her tattooing and tatters, and sweeps her long robes out of sight. Her mother, very like her in face, but gentler of mien, takes her place, producing two brass coffee-pots, with tops, one of which she sets in the ashes, the other upon the ground. To a man, who is our fellow-guest, and who at our entrance, lay asleep upon a mat at the back of the tent, is assigned the duty of roasting the coffee.

"THE SON OF THE SHEIK."

"He is a near relative of the family," says the interpreter, "and may therefore be honored by helping prepare for entertaining the visitors."

The son of the sheik, well-knit in figure and of good stature, but whose otherwise fine face is disfigured by weak eyes, now makes his appearance from another part of the village. His black-and-white abieh parts over an embroidered vest; his ample kafeyeh is also wrought in various colors, and one end is flung picturesquely across his breast. When the coffee is ready, he presents a cup to me upon one knee, and hopes I will honor him by tasting it. So princely is his deportment, so thoroughbred the grave courtesy of his behavior to the intruders upon his home, that I entreat David to convey him in his very best manner our appreciation of his kindness and our sense of our ill-deserts.

The acting host listens respectfully, his head slightly bent, and not a muscle of his visage relaxing from the pensiveness which seems habitual. Then he turns to me and makes a speech of several sentences.

"He says that in his father's absence, he hopes you will look upon him as the head of the household, and your host. That he is proud and honored by your visit to his poor hut. That he trusts you will remain long, and consider all that he has as yours," repeats David, conscientiously.

In response to our queries, the young man informs us that his father is sheik over about one hundred and twenty families, scattered over an area of five or six miles. The flocks and herds we have passed on the road since luncheon belong to the tribe.

"HANDSOME, BLACK-EYED AND MERRY."

The, until now, unsmiling countenance lights up as a little toddler of two and a half years runs into the tent and clasps him about the knees.

"My son!" he says pridefully, raising him in his arms. The old woman who made the coffee laughs sympathetically, and glances at me with the quick free-masonry of motherhood which is comprehended the whole world round.

"How many children have you?" I have David Jamal ask her.

"Two."

"Both are married, I suppose?" recollecting that she is mother-in-law to the sheik.

"Neither is married."

This "poser" is solved subsequently by the intelligence that her two bachelor sons are much younger than the favorite of the harem (who is not the young host's mother, by the way) and that her three daughters do not count in the computation of her "jewels."

There are twelve children, belonging to somebody—or somebodies—in the sheik's abode. Handsome, black-eyed and merry younglings, all of them, and as ragged and unkempt for the most part as they are jolly. We beg to be allowed to photograph a group drawn up in unconscious effectiveness across the open front of the "reception room," and it is pleasant to see the sheik's son persuading them to stand still, and holding his shy half-sister, the beauty of the band, by the hand to reassure her that we intend no harm. A second trial brings in more subjects, and in the foreground the baby boy, the apple of his father's eye.

"SHOWS BUT OBSCURELY."

They are a gentle, peaceable tribe, if we may trust the evidence of our observation among them this afternoon. Not a question is put to us, and there is not a symptom of surprise at our descent, horse and foot, upon the encampment. When we make a motion to go, we are urged to remain to dinner, and assured that we could be made comfortable for the night. Our diffident petition to be allowed to carry off pictures of the village and inhabitants meets with ready assent, the women (unaided by the men, we notice), eagerly looping back the western curtain to let in a stream of light upon an interior that, even thus, shows but obscurely upon the developed "film."

Good-byes have been exchanged, and our small party is in motion, when the sheik's son brings forward his boy—the fond smile which the baby alone has called to the serious face, irradiating it—and holds him up to kiss my hand. The little fellow, tutored by his parent, goes prettily through the form; I detach a bright flower I am wearing, from my dress, and drop it into the child's hand, and we carry away the image of the two in the foreground of our memory of the visit.

"THE TRIBAL POET."

A mile further upon our road we meet the tribal poet, and he is duly presented to our notice.

"Herein is novelty!" we can hardly await our parting to ejaculate. "A Bedouin bard! What does he do? and when? and how? and where?"

As a sequence, we enjoy at our tent door in the moonlight and succeeding the evening meal an intelligent and entertaining lecture upon Bedouin war-songs, love-songs, and horse-songs. To the latter, the Ishmaelitish muse most seriously and frequently inclines. A metrical translation of one copied from Alcides' note-book will give a fair idea of their scope and spirit:

I covet not the hoarded wealth the rich Damascene boasts;
I covet not the merchant's ships that trade with distant coasts;
But the man who owns the swiftest steed that sweeps the desert sands,
His wealth I covet more than all throughout the Prophet's lands.
I covet him, although he bears no saddle and no rein;
That horse I'll buy if he should cost all I e'er hope to gain.
Although my tribesmen call me mad—although they speak me ill,
I wish not more; my horse shall be my greatest treasure still.

The Bedouin war-lyrics are in the same key with the old border-ballads. While they may not rank in poetic merit with "Chevy Chase" and "Edom o'Gordon," we make notes of some which show fire of imagination, as well as of spirit. One, founded upon an actual occurrence, is worth relating here. The meagre outline can not impress the reader as the hearing moves us in the profound stillness of an evening as mild as May-time. The interpreter sits a few feet away, the moon full upon the dark, strong features, reciting one line at a time in the native Arabic, as in soliloquy, then rendering it into English for our enlightenment.

A young man's sweetheart was captured by a marauding band with whom her people were at war. All through the country-side went the call to arms against the robbers who had stolen the fairest treasure of the tribe. When all the men-of-war mustered before the tent of the sheik, he said there were too many. There would be no honor in attacking the enemy with an overwhelming force. He would take with him no boys, only warriors used to carry arms, and as a test, he decreed that none should follow him whose beards were not thick enough to hold the "beard comb" when stuck into them.

Now, the youth whose promised bride had been carried off was but twenty years of age and had no beard. Should he endure the disgrace of seeing the fighting men of the tribe depart to recover his betrothed, and revenge him upon her captors, and he be left behind with the women and children in the tents? So, he drove the beard comb into his chin, and leaving it sticking in the flesh, presented himself thus to the sheik. The chieftain was so delighted with his daring that he allowed him to ride forth at his side. The lover recovered his bride, and bore the honorable scar to his dying day.

Another song has to do with the adventures of Ali Diab, a sheik in the land of Moab, renowned for bravery. He was proscribed by the Government for daring violations of the law, and, while an outlaw, married one of the most beautiful women in Syria. Shortly after the wedding, he, his wife and a few horsemen were crossing the country when news was brought that a body of the pasha's soldiers were in close pursuit of them. The sheik ordered his wife to mount before him on his horse, that they might make a desperate flight while the way was yet open. She spat in his face and broke into biting reproaches. "I married you not for your beauty but your courage," she said. "I see you are a coward and no better than a woman."

The husband hastily detailed a handful of his men to remain with his wife; took the rest along "to guard his back," charged upon the soldiers, killed thirty of them, captured their horses; rode back to his bride and they escaped to the mountains together. This woman was so fair that a song was sung in her honor in Ali Diab's tribe, the refrain of which runs something like this: "We will beat the soldiers of the pasha, for the sake of the eyes of Ali Diab's wife."

The hero of this tale is still living in a hale old age, and is reputed to be the mightiest and wealthiest Bedouin in Syria.

RAILWAY STATION AT JERUSALEM.

CHAPTER XX.

"THE CITY OF THE GREAT KING."

OR a close, comprehensive view of the present city of Jerusalem and the encompassing hills, there is no better standing-point than that to which we have ascended within fifteen minutes after our arrival. The hour is close upon sunset, and this "loggia" upon the roof of the Grand New Hotel is an incomparable observatory from which to enjoy the rising and going down of the sun.

M. Antoine Gelat, the prince of hotel managers in kindly regard for the comfort of his guests, and a sort of intuition that anticipates desires and suits circumstances to their tastes, is our conductor. He says just enough to accustom us to our environment, then stands aside, ready to answer further questions, but volunteering no comment.

We lean upon the railing, our souls in our eyes, and gaze toward the Mount of Olives. What pilgrim does not first turn in that direction? The west is crimson behind it; the white road leading around it to Bethany is partly in the shadow creeping up the city-ward slope to the tall tower upon the top, said to mark the spot of the Ascension.

"Which cannot be,"—our host suggests, "since we read that 'He led them out as far as Bethany.'"

The intervening roofs and the outer wall of the city conceal Gethsemane from us. The evening light folds warmly about the dome of the Mosque of Omar, standing all these centuries as a type of the abomination of desolation, above the most Holy Place.

"Mount Moriah!" has been said in our ears—"and over there, Calvary, and where we stand, Mount Zion. Across the street is what is known as David's Tower. The foundation stones belong to the original structure. You can see where the more modern masonry begins."

Creeping plants grow rankly in the crevices between the massive stones; upon the flag-pole on the top, the crescent-and-star flap feebly in the sunset breeze; the shrill sweetness of bugle notes winds upwards above the noises of the street. A Turkish garrison occupies the ancient fortress, for long the citadel of the "city which David built."

"It is a little city now," we murmur, voicing the disappointment inevitable to those whose imaginations have drunk in since infancy, stories of Jerusalem the Golden.

"Beautiful for situation, the joy of the whole earth, is Mount Zion on the sides of the north, the city of the Great King."

The words roll like a majestic chant through the memory. The mountains are around about her still, but where are the towers we are bidden to count? Are her bulwarks worth marking? and her palaces—where are they?

"It was smaller still when David had reigned here thirty years," says one, "for Moriah was the Jebusite's threshing-floor."

Reluctant to admit what is making our hearts sink, we ask for the site of the

"A CITY THAT IS COMPACT TOGETHER."

Church of the Holy Sepulchre, and it is shown to us, almost beneath our feet. The Jaffa gate is close to the hotel; the street conducting from it is the narrow artery of city-life. Against a wall where passers-by must stumble over their feet, sit three men in attitudes of intense fatigue, or dejection.

"Professional beggars!" Mr. Gelat remarks. "Picturesque? Yes! picturesqueness is a branch of their business."

Nobody notices them, while we are looking on. This is the busiest thoroughfare in Jerusalem. We had not known that there were so many donkeys in the world as we see driven or ridden by here in one hour of the day. (In some parts of Europe, the ass is known as the "Jerusalem pony.") Some carry burdens

THE MOUNT CALLED OLIVET.

bigger than themselves of faggots, bales of merchandise, crockery, green vegetables—whatever supplies shops and private houses. Some are bestridden by turbaned countrymen, sitting so far back upon the hind quarters that the bulging abiehs conceal the animals' tails; upon others are women muffled up to the forehead, with children upon their laps, and panniers dependent from both sides. Red tarbooshes make spots of color everywhere; some of the wearers are in native costume, many in European; peasants are here in every variety of shabbiness possible to the striped robe of cheap camel's hair, that is graceful or ungainly according to the wearer's station and bearing; Syrian women in white izzars and colored mendeels jostle Jewish women with shawls over their heads, and uncovered faces; donkey-boys, fruit-selling boys, camel-boys trudging beside the ugly beasts lunging right through the heart of the crowd, their great spongy hoofs spattering the mud right and left from the sunken middle of the paved way; beggar-boys —even—and here we laugh,—newsboys, crying what look like evening papers,—are everywhere conspicuous. Some of the urchins sport the loyal tarboosh, most of them are bareheaded, and so far as we can judge at this height, not one has a personal acquaintance with a comb, or ever heard of a brush. It rained this morning, and from the mire and dirt the many feet stir up evil odors.

"THE PRINCE OF HOTEL MANAGERS."

"The joy of the whole earth!" Ah!

The lower rim of the sun kisses the hilly horizon, and through the mingling and nameless clamor of the lower world, a wild alien cry reaches our

MOUNT ZION AND THE MOUNT OF OLIVES FROM "POMPEY'S FIELD."

outlook, and turns us toward a square minaret, nigh unto the Church of the Holy Sepulchre.

Once heard, it is forever recognizable—that tremulous wail of the Muezzin, sent out five times daily to the faithful, whether they will hear, or whether they will forbear.

> Allah, hu akbâr!
> La illah illah Allâh;
> Siadnah Mohammed Râsoul Allâh
> Hayah Allah Il Salâh!
> Hayah Allah Il Fallâh!
>
> (God is great!
> There is but one God;
> Our Lord Mohammed is his apostle (or prophet).
> Come to prayers!
> Come to worship!)

Each phrase is said twice before passing to the next; at the conclusion of the call, the whole formula is repeated from first to last, faster than before, and to a different tune, if tune there be.

This is the famous call to prayer of which such capital is made by writers of travels and Oriental tales. It is, as I have said, peculiar. To be frank, it is not musical, and except when heard from a distance on a still evening not even pleasant. As to the devoutness with which it is enunciated and heard, candor compels the admission that the whole performance is perfunctory to the last degree, and that the effect upon the listener is very unlike what we have been led to expect from the use made of the "much speaking" of Moslems by Sunday-school orators and returned missionaries. As the many a's (as broad as circumflex accents can make them) are shaken and quavered and trilled over the red tarbooshed heads of the faithful, we seek in vain for tokens of the profound impression we have been told is invariably produced by the solemn summons.

Between the machicolations of the venerable tower opposite, are visible soldierly figures, lounging and smoking and evidently off-guard. So far from kneeling with their faces toward Mecca and going through the form of prayer, they continue to saunter, to chat and idly survey the street passengers, who for their part, are as deaf to the worse than vain repetitions agitating the air about them as if there were not a minaret within ten miles of Jerusalem.

The beggars pull themselves up and shuffle out of sight, the sunshine having left them in chill shadow; camel-drivers hold on their heedless way in the middle of the street; the driver of a muddy hack drawn up in a corner near the Jaffa gate, eyes with more sense of humor than one usually detects among the saturnine people, a lively tussle between two donkey-boys in front of a huckster's booth; the lordly "kervasse" (orderly) of a foreign consul—gorgeous in gold lace, portentous as to sword, official baton and ultra-official frown—awaits his chief's return from the photograph and olive wood bazaar across the street. Nobody prays that we can discover, unless it be a white-bearded man upon a neighboring house-top, sitting motionless with his back toward us—but, as we now perceive, with his face turned away from Mecca.

We are unaffectedly sorry to take the pith out of an illustration so potent in stirring up Christian listeners to emulate the devotion prevalent in "lands be-

"PICTURESQUENESS IS A PART OF THEIR BUSINESS."

nighted"—but the facts are precisely as I state them. Four minarets rear balconied heads heavenwards within range of attentive ears, and a good glass shows us as many muezzins popping out of inner chambers, like cuckoos when the clocks strike the hour, and uttering, still cuckoo-like, their "wobbling" cry.

"We expected to see every man fall upon his face and go through his prayers," we complain. "It must cost a great deal to build up these minarets—(there must be at least a dozen in this little city)—and to employ the muezzins, yet nobody pays any attention to them. Is it so everywhere!"

TOWER OF DAVID.

A Jerusalem Roman Catholic comes forward to reply:

"Even in the East, people are too busy in towns to pray so many times a day, unless they are very religious. Most Moslems here content themselves with morning and evening exercises. In the country it is different."

We face the ruddy glory of the west, and without the exchange of a word or glance, each knows that the other recalls "The Angelus," and the bowed peasants in the foreground, the prayerful stillness that may be felt of the fields at eventide.

"THE DRIVER OF A MUDDY HACK."

Drawn darkly against the sky that palpitates with fervid color, is a windmill solitary upon a distant hill; pale stone buildings, some large and all new, start into prominence in middle distance and foreground. The city without the gates is growing fast, and will soon exceed in dimensions, as in neatness and stability, the old town with its strait, steep alleys and irregular lines of houses.

There are no gas-works and no water-works. After nightfall, if one would venture abroad, he must have an attendant with a lantern. The water supply is drawn from cisterns of vast depth, some of which never go dry. In a drizzling rain the streets are foul with semi-liquid filth; when the showers fall heavily, much of this is swept down to lower levels. One does not care to calculate what proportion of noxious matter percolates through the soil to the mighty honeycomb of cisterns underlying the town.

Old Jerusalem is buried many feet below the surface of modern by the rubbish of seventeen overthrows of the City of the Great King. All her pleasant places, often restored, were laid waste beyond the hope of rehabilitation, when the plowshare of Titus furrowed the hill of Moriah before the generation had passed away that heard the Galilean Teacher's lament over the "Jerusalem that killed the prophets." The Jerusalem who in three days' time would hound to a malefactor's death the Greatest of them all. To our left, between us and the long, straight brow of Scopus and rising above the city wall, low tombstones crown another hill. Our eyes never leave it for long, however they may stray as convent, hospice, asylum and church are named to us. The horizontal sun-rays strike and linger upon it, while we take one final look before going below, and the light brings out the tender hues of verdure that always clothe it, even in winter.

> The green hill far away,
> Without a city wall—

is what, after all, we have come out to see in this far land—a veritable wilderness to him whose happiness depends upon the luxuries of nineteenth century civilization.

Voices—a heavy bass and a light treble—clash together between the walls shutting in the stairway by which we have climbed to the roof. A dumpy woman pants to the upper floor, followed by a thin man. We know the smooth, round face of one and the sharpened red beard of the other. They are not staying in the Grand, and have only come to get the view. Mr. Gelat advances urbanely; Dr. Sharpe introduces, first himself, and then his wife, and "hopes there is no intrusion."

The hospitable manager sets his mind at rest in some fitly chosen words.

"You are well known to me by reputation, Dr. Sharpe, and I shall be honored by any use you can make of my house or my poor services. The view is, as you see, extensive. You are, I presume, familiar with the several points of interest visible from here?"

Mrs. Sharpe utters a little cry. She has walked to the farthest end of the "loggia," and clasps her hands ecstatically:

"I had *no* idea you were so near that dear, *dear* Church of the Holy Sepulchre! How lovely!"

"A misnomer, which perpetuates the wildest fable of the hundreds the Middle Ages have bequeathed to us!" we hear her husband begin—and hasten down the steps.

THE JERUSALEM OF TO-DAY.

CHAPTER XXI.

"THE MOUNT CALLED OLIVET."

IT is still made conspicuous among the mountains round about Jerusalem by the olive groves clothing the sides and summit. In spring-time, the apparently barren reaches which are unshaded by trees will be mantled with verdure and flowers. In late autumn, and in the winter, the mighty slopes are a study in neutral tints, the silvery gray of the olives contrasting but faintly with the darker uniformity of color in soil and stones.

If we would ascend to the church and convent built upon the crown of the long hill, we must walk or ride. Driving in any kind of vehicle is out of the question. This has been set fairly before us, but we sally forth from the Jaffa gate, with an inadequate conception of what lies before us on this bland December afternoon. Massoud, refreshed by a couple of days in the stable, curvets and prances ahead of the two donkeys ridden by David and his charge. Jamal's is a brisk brown beast, with an amiable, conscientious amble, a pigmy beside the dignified creature, white as wool, allotted to me.

"Tradition tells us that the Queen of Sheba rode a white ass when she came to visit Solomon, madame," I hear, as we skirt the city wall and see upon our left the heights where Titus massed the Roman legions for the terrible siege begun at the Feast of the Passover, A. D. 70, to end five months later in the destruction of temple and citadel. "And kings' sons rode upon such. It is therefore counted a royal animal."

The royal animal beneath me is broad and stalwart and comely—of his kind. He is also so lazy as to require the jerk of Serkeese at his head, and the goad of a second boy upon his flanks when we begin to climb. The flesh may be strong, but the spirit it encases is at once unstable and rebellious.

I have not found this out when we pace down the long declivity, passing on one hand the "green hill," on the other the Damascus gate. The road is wide and smooth, and near the bottom bears away from the city wall to cross a bridge.

"The Brook Kidron!" says the guide.

We draw up to the side of the bridge and look down into the dry valley. It is not even a ravine now, which spring floods might fill. Olive trees grow upon the accumulated debris of eighteen centuries. If there be water in the ancient channel, it never makes its way to the surface. The so-called Tomb of the Virgin

(SO-CALLED) TOMB OF THE KINGS, WITHOUT THE WALLS OF JERUSALEM.

is on our left hand. In a few months, upon the day fixed by ecclesiastical dogma as the birthday of our Lord's mother, thousands of pilgrims will pitch tents and booths about tomb and chapel; brass bands will be in full blast all day and all night long; all manner of merchandise will be peddled along the road.

"Just like a picnic," says a Greek Christian whom we encounter on the bridge.

The monument with a square base and peaked tower in the deeper part of the valley upon our right, is known as the Tomb of Absalom.

"WE GAIN AN OPEN SPACE."

"Where he was never buried," comes from the back of the brisk brown donkey. "If the Captain will look at second Samuel, eighteenth chapter and seventeenth verse, he will read that the young man Absalom was cast into a pit in the wood of Ephraim. Learned men think that this may be the pillar alluded to in the eighteenth verse (if you please, sir!) which 'Absalom in his life-time reared up for himself in the king's dale.'"

At the name of the undutiful son a pathetic scene rises before us. We turn silently to another part of the sacred record and read how "all the country wept

POOL OF SILOAM.

with a loud voice" as the deposed king and his followers passed over Kidron; then,--that:

"David went up by the ascent of Mount Olivet, and wept as he went up and had his head covered, and he went barefoot; and all the people that was with him covered every man his head and they went up, weeping as they went up."

We are still reading when rasping voices smite our ears. Close to our stirrups is the vanguard of the forces drawn up on the sunny side of the walled way by which we are to make the ascent. The village of Siloam, the houses of which are scarcely distinguishable from the gray rocks to which they cling, lies a short half mile away. Above and beyond it, is the group of wretched huts tenanted by Jerusalem lepers, and this thoroughfare along the Kidron, and so on into the open country, is the pariah's stamping-ground. Old and young of both sexes haunt the spot on every moderately fine day. Some carry pails and baskets into which passing market-gardeners drop vegetables and fruit. One and all beg, and that continually, shamelessly making capital of mutilation and other disfigurement that sicken our souls and senses. David charges into the line closing across our way:

"We went this way this morning, and gave you backshish—and you know it!" he vociferates in stinging Arabic. "I am ashamed of you!"

At the admonition they slink back mute, if not abashed. That a party must not be bled twice in one day is a tenet in their unwritten code of honor and business.

The narrow roadway soon twists and contracts into a bridle-path. A path by courtesy only, for the rolling stones of all shapes and sizes are as plentiful under our feet as on both sides of the route. Serkeese tugs hard at my royal beast's halter, the other boy applies a vigorous shoulder to the creature's haunches as we go up and up, tacking across the steep breast of the hill. Now and then, the efforts of both drivers are ineffectual to prevent the donkey from stopping short in an especially precipitous place, and, drawing a long breath that heaves one high in the saddle, venting his emotion in what is surely the most ear-splitting and incredibly protracted bray that ever expanded asinine ribs. The experience is exasperating—but funny—up to the seventh repetition.

Still tacking, we gain an open space,—a waste of loose stones—not far from the Church of the Ascension, and turn for a long look at the view. The day is very still; the sunlight is colorless; beyond gray hillside and melancholy olive-trees, and across the valley where the Kidron flows no longer, and cool Siloam is choked by the heaps upon heaps of pleasant places laid waste—rise the hoary walls of Jerusalem.

Here Hushai overtook David, and was sent back into Jerusalem by his master to defeat the counsels of Ahithophel. For aught that we, or any one now alive can know, it might have been upon this very spot that our Lord sat down to rest, after departing from the temple, and staying in the outer vestibule at His disciples' appeal to look upon the great stones to be thrown down before the passing of the faithless generation. The leaves of our "guide-book" do not flutter in the waveless air as we look for the story:

"And as He sat upon the Mount of Olives, the disciples came unto Him privately, saying: 'Tell us, when shall these things be? And what shall be the sign of the coming?'"

If this be true, here, or near where we read these words, were given the

"THEY SLINK BACK MUTE."

parables of the Ten Virgins and the Ten Talents, and the awful picture was drawn of the division of the righteous from the wicked.

We visit Gethsemane on the way down. The inclosed "Garden," never large, has been subdivided into what may be called flower-plots, that Greek and Roman Catholic may no longer disgrace the name and memory of a common Master by hideous wranglings. Each sect has its own paled-in territory. Having left our beasts of burden and our attendants at the end of the short lane leading to the main road, we have the place to ourselves, but for a lay-brother in a wide felt hat who is "pottering" with a hoe among the borders of old-fashioned cottage-flowers, under the most aged of the olive trees. Without so much as visiting the Grotto of the Agony, or the connecting subterranean chapel where are shown the tomb of St. Anna, the mother of the Virgin Mary and that of Mary's husband, Joseph, or caring to measure the distance—"as it were a stone's cast"—dividing the Grotto of the Agony from the place without the garden—designated as that where Peter, James and John slept while their Lord wept and prayed—we bow to

"OVER AND BEYOND THE ROUNDED CROWNS OF THE OLIVES."

the solemn influences of the hallowed precincts. Somewhere on this sunny declivity was the retreat to which the Man of Sorrows so often resorted with His disciples, that the arch-traitor knew only too well where He was to be found after the Passover feast. The olives, gaunt and gnarled with years, that gleam in the afternoon light, have sprung from the soil that gave foothold to the trees of the Gethsemane He loved. He must have crossed the brook by some such route as we have taken, and the tinkle of the running water may have joined with the sigh of the night-wind in the olive-boughs to lull the overwrought disciples to slumber. We forget traditions and monkish superstition in pondering these things, our faces

AGED OLIVE-TREE IN GARDEN OF GETHSEMANE. (203)

GARDEN OF GETHSEMANE, THE HOLY LAND.

turned toward Jerusalem that, held not back, after slaying the prophets, from denying the Holy One and Just, and killing the Prince of Life.

The muezzins are answering one another from minaret to minaret, the harsh cry mellowed in the passage across the valley, when we withdraw our eyes from what our hearts and memories must ever hold. David stands without the gate, and the lay-brother comes up with a sprig of rosemary.

Seeing me examine the gift in the hope of finding a woody stem, he goes back into the garden for a stouter spray, and breaks off a piece of lavender as well.

"Both will grow, if planted," the interpreter says for him to me.

"Would he give me earth in which to plant them?"

He not only will, but hastens to transfer a generous double-handful to a stout envelope David produces from his inexhaustible pockets, and furthermore insists upon my acceptance of marigold and princess-feather seeds, pulled from dry stalks beside him.

David smiles, gravely indulgent to the simply uttered admonition accompanying the humble marigold.

"He warns you, madame, that the flower requires sun and a warm soil. He thinks you are from a far-off and a cold country, like Russia, from which hundreds of pilgrims come yearly to Gethsemane."

It falls upon our exalted mood like the touch of a warning hand that we will never return from our "far-off" home upon a second pilgrimage to the Garden of the Agony and Bloody Sweat. We have repeatedly remarked to one another that full appreciation and right feeling do not come to us until we review the moment or hour in thought—that warmth and thrill then are like the afterglow of the departed sun. We shall always be glad that the Figure, never distant from our spiritual apprehension in our daily walks and rides about Jerusalem, has never been nearer than in the solemn shadows now gathering under the trees of the garden into which He entered that night with His disciples, "As He was wont." The breath of His presence, the calm, and the blessing of it may well abide here now and forevermore.

We ride a little way back up the Mount called Olivet, to possess ourselves of a glimpse we had awhile ago, over and between the rounded crowns of the olives, of a hill-top rising dimly through the gauzy haze, "beyond the city wall."

"*By thine Agony and Bloody Sweat, by thy Cross and Passion, good Lord, deliver us!*"

CHAPTER XXII.

BETHANY.

AN excellent road connects Jerusalem with Bethany. The city boasts of but one handsome carriage ("the Landau," belonging to the Grand Hotel), yet comfortable conveyances may be hired without the Jaffa Gate. That into which David packs our rugs, the camera, and lastly the women of the little party, is commodious, if shabby.

"As you will observe, madam, it has been washed this morning," says the honest dragoman, withdrawing a space to contemplate complacently the effect of

"IT HAS BEEN WASHED THIS MORNING."

the unwonted process. "As I have said to the driver, a muddy carriage does not go well with a day like this."

We have alighted to look back at the city, and to remark again upon the

MODERN BETHANY.

growth of the settlement without the walls, for the most part hospitals, convents and the private residences of consuls and well-to-do citizens. The day is glorious; respiration is a luxury. The stony hill-sides can never look cheerful except in the brief blossom season; but under the cloudless blue arch of this sky, they are resigned in expression and almost benignant.

My companions within the lately-washen vehicle are a newly arrived fellow-country-woman, bright of face and of nature, and the native Syrian matron who acts as my intelligent interpreter in the jealously-guarded homes where the foot of a strange man may not tread. We have crossed the waterless bed of Kidron; run the gauntlet of the leprous beggars, mustering in force on the south side of the wall of Gethsemane, holding our breaths in passing, the odor of the collective bodies being overpowering when so large a delegation is abroad; have traced with serious eyes the "Path of the Captivity" winding around the shoulder of the height upon which the city is built, said by tradition to be the way along which Our Lord was led after his betrayal.

We have flanked Olivet and come into sight of the open country. The road is lively with foot-passengers, generally market-women with great baskets or bundles of green stuff upon their heads—oftener bundles than baskets. It is surprising what variety of commodities can be contained in a fold of the Syrian woman's blue cotton robe, or be enveloped in the dingy cloth knotted about her cargo. Beets, turnips, cauliflower, carrots, eggs, oranges, potatoes, cheeses—chiefest of all, cabbages—are thus carried with ease and in safety. Donkeys pace by, under loads of faggots and greens, and dark-visaged men seated stolidly upon the beasts' rumps, and we meet a string of camels, coming up from Jericho, loaded with oranges and citrons, and hugging one side of the white road. Upon the hill-top to the right we see a building which we are told is the reputed house of Simon the Leper.

"Then we must be nearing Bethany?"

"We are in Bethany!"

The carriage has come to a full stop again, and we get out in blank bewilderment. Upon one hand the ground slopes away into stony fields; upon the other, half ruined huts of stone, perhaps twenty in number, straggle up a rudely-terraced hill. The invariable group of children, a couple of beggars, and divers women, appear in doors and from behind walls to stare at us. The terraces and houses have an intermittent backing of the saddest-looking olives we have yet beheld in this saddened land.

"Bethany!" we reiterate in wonder and disbelief. "Where is the town?"

This is all of it, and the two women who are bribed with a few "metalliks" (pennies) to stand out upon the platform of their house to be photographed are representative matrons. They have no scruples about letting us see their faces, etiquette in this respect being less stringent among the peasants than in the town and with the higher classes of society. Whatever other objects and occasions of carefulness and trouble these housewives may have, the business of keeping themselves and their children clean is not ranked as an essential to peace of mind or comfort. Their own gowns are ragged and dirty; their hands horny and filthy, albeit the nails, like those of the little girl who lugs an inconceivably dirty baby upon her hip, are dyed with henna. The children are, as the uninitiated new arrival at my elbow shudderingly whispers, "horrors!" sore-eyed, barefooted and unclean. One, sour of visage, a kafeyeh, stiff with dirt, bound above matted locks, hugs a mangy dog, whose eyes are as sore as his master's.

"YOUNG DAVID."

We are on our way up a narrow, steep alley, abominable for pollutions, but the only means of approach to "the best house in Bethany," when a boy bounds into the track, and stops to inspect us. By some accident, his face is clean, and the soiled sheepskin jacket beneath his abieh does not disguise a certain careless grace of bearing. His eyes are bright and healthy, his cheeks glow rosily.

"'Ruddy, and withal of a beautiful countenance, and goodly to look at!'" quotes Alcides "Stand, young David! and let me have your picture."

The model frowns slightly, the sun striking full in his eyes, but the erect carriage of figure and head that gives something kingly to the boyish figure is preserved.

"Such a fellow must be regarded as a freak in Bethany," comments his admirer, clicking the film into position.

The best house in Bethany is at hand. The stone platform, or terrace, on which it perches sets the front and only door ten feet above the level of the nearest neighbor's. A ladder, fitted with rungs, not steps, must be mounted if we would accept the invitation extended in smile and beck by the mistress of the eyrie. One by one, we three women adventure the climb, and stand beside our hostess on a ledge, maybe fifteen feet in width. It looks down into the court of the "House of Lazarus," a blackened building that must now be used as a mill or granary, for through sashless windows protrude sheaves of grain and bundles of straw.

In the reeking court-yard where the sun seldom falls, two sparrows twitter happily over some kernels that have dropped from above. Another pair, as busy and cheery, sport upon the broken stones of the time-stained wall, pecking their noon meal out of the chinks and crannies.

"'Are not two sparrows sold for a farthing,'" one of us breathes softly before leaving the little preachers and the cheerful light of day behind.

The door opens into a vaulted chamber that cannot be over twelve feet square. There is no window, but six queer niches in the sides of the room are utilized as cupboards. The walls are whitewashed, and above the niches are gaudily-colored designs of palms, flowers, fruits, and odd green *hands*, with outspread fingers pointing upward.

"To keep off the Evil Eye," explains my interpreter, avoiding looking at the mystic signs while she speaks.

To direct attention to them by any manifestation of curiosity on our part would excite suspicious alarm.

Quilts and matting are dragged from a pile at the back of the room to supply us with seats. A man, his wife, and five daughters of assorted ages, have their only home here, sleeping, eating, sitting and living here the year around. Cooking is done over an earthenware brazier, exhibited to us pridefully by the housewife. The bread, she tells us, is kneaded into flat rounds which are laid upon hot pebbles in the brazier, patted down, and covered with a fire of dried camel's manure. A heap of this fuel fills one cupboard! Flour for each day's provision is ground in the morning. A laughing-eyed girl of thirteen, persuaded by the gracious arts of our interpreter, sits down upon the floor, takes the mill—a stone, laid in a socket and worked back and forth by an upright pin, or handle of wood—between her feet, pours in grain and begins the work.

"But it is nothing without the song," interrupts the interpreter. "Sing, will you not? These ladies have never heard the Syrian grinding-song."

The fair performer requires as much coaxing as the musical prodigy of an American provincial circle, but demur and diffidence finally overcome by the winsome interpreter, she chants in a fresh, tuneful voice, to the accompaniment of the mill. The effect is marvelously melodious, the hum of the stone in its groove keeping perfect time with the song. Our Syrian matron translates what I seize a moment to jot down in my note-book, and subsequently pass over to Alcides the Rhymer. I wish it were possible to give with it the beat and whirr that, as I read the faithful translation which retains the very measure of the rude ballad, seems to underrun the voice in my ears:

> Soldiers! soldiers! far over the sea!
> As we grind the yellow grain, may the Prophet hear our plea!
> May he spread his mighty arms in protection round your host,
> And hide you in their shelter safe when perils threaten most.
> May he guard his faithful followers wherever they may be!
> Soldiers! soldiers! far over the sea!

The cupboards, unscreened by curtain or door, hold, besides the cakes of fuel, earthenware jars and bowls, the inevitable apparatus for coffee-making, and a big metal pan covered with a mat of plaited straw. The day's baking is in this basin, and since we cannot wait for coffee, we must eat, each of us, a morsel of the tough, warm cake. It is curiously stamped by the hot pebbles, and dark-gray in color. Yet we accept it with smiling gratitude, and feign to swallow it, retiring in fair order toward the door, while our interpreter voices our thanks for hospitality received.

Jamal and Alcides in the lower court are the cynosure of an admiring group. The Bethany women are notable of their kind, carrying on (for Palestine) a thriving trade with Jerusalem in butter milk, eggs, cheese and butter. One of the most prosperous listens with downcast eyes to David's talk of crops and weather, cunningly drawn out while Alcides retires to a ledge far enough away to focus the modern Martha of Bethany. Every line of her coarse features expresses the cumbering cares of much serving in home and market. Her baby sleeps in a stout bag suspended between her shoulders by a cord passed across her forehead; on her head is the flat basket that went with her and the baby to town (a Sabbath day's journey), filled with butter, cheese and eggs, and holds now sundry parcels of groceries and other household necessaries. The baby-bag is a curious thing, made of dyed goat's hair cords, ingeniously netted and trimmed with long fringes.

"Will she sell it?" I ask.

"She will sell anything!" tersely, and the bargain begins.

Other and shrewder lines cross Martha's visage. She is in her element.

Withdrawn a few paces from the battlefield, we are kept by the laughing interpreter in touch with the principal actors. Martha draws the first blood by a firm demand for sixteen shillings, English money (four dollars).

"Is the woman mad!" David raises appealing eyes and hands to the presumably sane spectators. "We don't want to buy the baby, too! Only the bag!"

Martha does not join in the merriment excited by the wilful misinterpretation. Buying and selling are not matters for jest. She chaffers valiantly, but the end of the ten-minute duel is the purchase of the gayly-colored netting at half the sum she asked at first. My Syrian friend interferes when I order the coachman to put the bag with other belongings of ours in front of the carriage.

"LISTENS WITH DOWNCAST EYES."

"It will go first to my house, if you please," she says, significantly. "I have a garden where it can be sunned, and fumigated with burning sulphur, and so be made quite safe to be packed with clean things. You shall have it in a few days."

A warm discussion arises on the way home. Alcides has been thinking.

"We see how little social and domestic customs have changed in this part of the world since Bible times"—is the summing up of these meditations.

"The Bedouin of 1894 A. D. is the Bedouin of 1894 B. C. That was in Abraham's day—wasn't it, David?"

"Between the eighteenth and twenty-second chapters of Genesis, Captain."

"Dress, modes of speech, tricks of trade, marrying and giving in marriage—all these things are so much the same now as then, that I believe the original Martha of Bethany lived and looked very much as her sister of the nineteenth century lives and acts and looks now. The whole family probably lived in one ten-by-ten room, and sat on the floor to eat and talk, and lay upon dirty mats to sleep, and crouched, on rainy days, over a brazier in a windowless chamber, and were upon as bad terms with soap, water and towels, and combs, as at present."

It is a bomb that bursts disastrously. The sketch is repulsive, unnatural—we declaim—if not actually blasphemous. We remind the transgressor that the bleak hills about us blossomed as the rose in the days of Israel's glory; that

Herod the Magnificent was on the throne and abundant in public works when the Nazarene Teacher withdrew to Bethany at eventide after the day's disputation in the Temple, or many hours of healing and talking in market-place and by the waysides. At last the gentle voice of the sweet-faced American traveler is heard. She speaks modestly—but confidently.

"I cannot think that our Martha was like that! Or that her house and

"IN BETHANY."

clothes—and children, if she had any—were dirty. For you know—'Jesus *loved* Martha, and her sister, and Lazarus!' And Bethany was His chosen resting-place. Martha's home may have been humble. It must have been neat and peaceful."

A sudden and a lasting calm ensues that has not been broken when we quit the carriage to walk a little way up the side of Olivet to the old Bethany road, and gaze down upon Jerusalem as our Lord beheld it four days before the shout of "Hosanna" was exchanged for the howl of "Crucify Him!"

CHAPTER XXIII.

ON THE HOUSE-TOP.

THE roof is gained by means of a succession of narrow stone stairways built upon the outside of the house. At the head of the last flight the host meets us, having heard our voices.

A handsome, well-made man, in European costume except for the red tarboosh that proclaims him a Turkish subject, he smiles genially at our approach.

"You are welcome," he assures us in excellent English. "I am sorry my wife is not at home, but she will return soon. Will you walk in?"

"Presently," we reply.

For a moment we must get our breath and look about us at the odd, charming place to which we have climbed, and fill our lungs with the delicious air, the sweetest we have tasted in Jerusalem. It pours in upon the enclosed house-top through geometrical patterns of short tiles set horizontally in mortar upon a stone wall six feet high. In the middle of the area, blocks of stone are built into a square platform with sloping sides, and this is surmounted by a box filled with earth. Flowers grow luxuriantly on all sides; pots of flowering plants and vines stand against the sunniest side of the little courtyard and crowd a graded wooden frame up to the top of the wall. It is the bonniest nook our eyes have lighted upon since we left Beirût gardens, and we say it in frank admiration. Gratified by our appreciation of the advantages of his dwelling-place, the host enlarges further upon these.

"We are raised so high above the level of the city that we get no dust, no fog, no bad smells, no noise. And the view—ah! that you shall see when we go into the room. Here it is only to be had in bits, by peeping through these holes. You see this is one of the oldest houses in Jerusalem—very old and solid. The walls are thick, as you have noticed. It must be hundreds,—yes, hundreds of years old. At that time the Moslems had entire possession of the city, and built their houses to please themselves. Their wives must have air and sunshine, and being women—and human beings"—smilingly—"they would not be contented without seeing something of the world. So they made these walls high, that they could not look over, and that nobody could look in at them, and exchange signals, you know, but with peep-holes out of which the women could see what was going on outside."

"What does your wife say to this sort of prison?"

"She does not mind. It makes the place more, private; in fact, gives us a pleasant sitting-room when the sun is not too hot. And on moonlight evenings when the weather is mild and the flowers in bloom, a man cannot wish for a nicer place for smoking his cigarette with his wife by his side and maybe, friends dropping in to help us enjoy it."

"You do not hide your wife away then behind curtains and lattices after the fashion of some of your countrymen?"

"I *trust* my wife!" drawing himself up unconsciously, yet with a good-

"AN ODD, CHARMING PLACE."

humored laugh. "We of the Greek Church have generally more liberal views on some subjects than Moslems. Ah?"

We have heard nothing except the soft rush of the breeze through the "peep-holes" and the light rustle of the vine-leaves against the wall, but he turns expectantly toward the staircase.

Something white rises into sight in the doorway giving upon the flight, and a little woman enveloped from head to foot in an izzar, trips forward upon the stone floor, pauses abruptly at perceiving us, and raising the blue mendeel from her face, stands hesitatingly, holding nervously to the edges of the white mantle which is usually cast off the instant the wearer finds herself within doors.

"My wife, madame! my wife, sir!" utters the proud husband. She kisses my hand, lays hers timidly into that extended by the masculine visitor, and looks in mute appeal to her mentor.

"THE HOST MEETS US."

"She understands English when it is spoken to her, but she will not try to speak it herself," he says, indulgently. "She would say that she does not comprehend it, but I know better, and I am sure, too, that she could speak it if she were not afraid to attempt it. She is what you call diffident."

At a word in Arabic, the little woman motions us to enter her dwelling, stands aside until her husband has passed in after us, slips off a pair of low shoes, with coquettish rosettes upon the instep, at the threshold, and follows us in her stocking-feet. They are pretty feet, we notice, and clad in scarlet stockings, and she looks very small for the loss of her slipper-heels.

"Let us make a little chamber, I pray thee, on the wall, and let us set for him there a bed, and a table, and a stool and a candlestick; and it shall be, when he cometh to us, that he shall turn in thither."

We comprehend it all with one glance around the "little chamber." How the prophet might gain the retreat thus prepared for him without entering any other part of the house, and without speaking to any of the family. How welcome, too, was the absolute quiet of the seclusion thus offered the foot-weary, nerve-tired man of God.

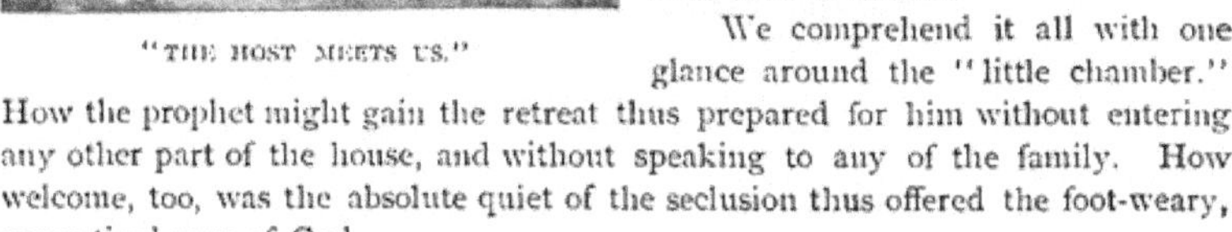

The room is of fair dimensions and massive in construction, with a groined roof. The windows are deeply embrasured, and from one end projects an alcoved casement, raised two feet or more from the floor, and cushioned with red chintz, as is the conventional divan running around two sides of the apartment. Before we sit down we are invited to look at the view, and for this purpose we are told to step, although we protest, into the cushioned alcove. It hangs upon the outer wall, as we now perceive, and overlooks, beyond the nearer jumble of tiled roofs and chimneys, the Mosque of Omar, about which we can trace from this altitude the boundaries of the Temple Area. A sudden dip outside the city walls means the Valley of Jehoshaphat, and then arises the Mount that is called Olivet, with the church and campanile upon its level brow. Scopus joins it on our left. If the Jew householder of A. D. 70 built as high as his Moslem successor, what an observatory would this have been from which to mark the manœuvres of the investing Romans! The free air sweeps gladly by us, the tranquil blue of the Syrian sky, falls, an unruffled canopy, behind farther and bare hills. Titus leveled all the trees about the city, and Palestine has never stood firmly enough and long enough upon her feet since to clothe her desolate places with forests.

"ONE CANNOT WISH FOR A NICER PLACE."

The floor of the house-top chamber is covered with Damascus rugs, and besides the divan, there are several light chairs; cupboards are numerous, and hidden by curtains. One is exposed because it is filled with books. There are little stands here and there, and bits of bric-a-brac, for the master of the quaint abode is a popular dragoman and has traveled much. The wife, black-eyed and "sonsie," wears a blue cloth bodice and skirt, a gold chain and an embroidered mendeel, and a bouquet of living flowers is pinned to the bodice. She is as neat and bright as her home, and steps noiselessly about making ready for our entertainment. A stand, bearing small glass dishes of quince conserves, and a basket filled with candies and nuts, is set before us, and upon another is arranged a tea-equipage. We have been given the choice between tea and coffee, and have gratefully chosen the former. The little matron opens a blind door in the wall, disclosing spirit-lamp and kettle, and falls to work to make the beverage my soul loveth, deftly and swiftly.

We have sipped it satisfiedly and tasted the conserves, and munched a bon-bon, and the hostess is outside in the elevated courtyard, talking with a neighbor who has called on a matter of housewifely business, when we draw from her husband the idyl of the house-top. It comes about through our comment upon the circumstances that the izzar—a cumbrous and not becoming garment, being, as I think I have already said, merely a square sheet draped and tucked and banded about the figure—is worn by women of all nationalities and religions in most cities of Northern and Southern Syria. It is a Moslem custom, we opine, and marvel at the adoption of it by Christians.

"It happened in this way, you see," is the next step to the story we have set our hearts upon getting. "Many, many years ago, as I told you, the Christians—Greeks and Roman Catholics alike—were accounted as nothing by their masters. I have heard my father say that in his boyhood it was not uncommon for a Turkish soldier, in passing a native Christian, to kick off his own slipper into the middle of the street, and say, 'You infidel dog! go get my slipper and put it on my foot!' The Christians did not care to muffle up their women and hide their faces, but they were forced to do as the Moslem women did, to save their wives and daughters from insult when they went into the streets. So they fell into Moslem fashions in this and other things, and even now must wear the izzar and the mendeel in public."

The little wife still chats with the neighbor, the Syrian sunshine bringing out the gleam of every link of the gold chain, and blazing back from the broad golden bands upon her wrists. She looks so prosperous and happy that the question follows naturally upon our glance at the fair picture:

"How long have you been married?"

"Nine years."

"Is it possible!" for the wife does not look to be twenty. "She must have been a mere child at the time?"

"She was sixteen. But"—a half-embarrassed laugh—"I made up my mind that she should be my wife the first moment I set eyes on her—a girl of thirteen, and just home from school."

This is promising—and refreshing in a country where marriage is a synonym for barter and sale.

"Ah! love at first sight?"

He looks doubtful.

"As to that, madame, we do not think so much of love here as you do in your country. It is more a matter of business."

To prove the assertion he narrates how, having arrived late at night at his sister's house after a protracted tour with a party of English people, he had slept late, and awaking at noon, espied through the window a young girl passing back and forth to a well in the court-yard. How, after watching her for an hour or so, he had arisen, and breakfasted, and while at breakfast, asked his sister the name of the stranger.

"THE LITTLE WIFE STILL CHATS WITH THE NEIGHBOR."

"This was one o'clock in the day. By six that evening my sister had made application to the girl's parents for their daughter in my name."

The father insisted upon a truce of three months, and the lover waited, voluntarily, six, before renewing his suit.

"I told him that my mind had not changed; that it would never change, but that if he said 'Wait six years,' instead of six months, six years it should be. I was given leave to visit at the house twice a week, and talk with her father and mother."

"Not with her?"

"Oh, for that matter, customs have altered with us since the times when a woman was not allowed to look at her betrothed until she was his wife. When I

(220) ROBINSON'S ARCH, JERUSALEM.

went in, she did not run out of the room, as was the rule in her mother's young days. I would say 'Good evening,' and she answered 'Good evening,' when I went in and 'Good night' when I went away. And now and then I pretended to be thirsty that I might ask her to get me a glass of water,"—a little shame-faced at the confession. "Further than this we never spoke together. It was three years after I first saw her before we were married. No! I had never seen her alone for one minute. I respected her too much to get her talked about. As it was, many thought our behavior very strange—quite a new fashion. They said it happened through my knowing so many Europeans. The wedding, too, caused great talk. Up to that time everybody was married at night, and it was the custom with us Greeks for the bride to stand just inside the door to receive the guests and to hold out her hand for whatever money, or other present had been brought to her. I said:

"My wife shall not play the beggar, and we will be married at noon and invite whoever will to come in the afternoon and wish us well, and have some refreshment, and then go quietly away. There will be no drinking and feasting and singing and dancing all night and all day, such as is the custom with others, for maybe two or three days together. And so it was. And many have followed our example. The Patriarch of Jerusalem has expressed himself as greatly in favor of the new order of things."

"You served for her three years and six months,—half as long as Jacob for Rachel," I remark, a little mischievously, as the wife, again leaving her sandals at the door, comes in to stand behind her husband's chair, laying her hand upon his shoulder.

"Yet you do not think so much of her as we do, you would have us believe! And you told us just now, that it would be accounted shameful for a woman to allow herself to love a man before she is married. I call yours a love match out-and-out, on your part, and, I more than half suspect, on your wife's also. Don't you believe she had learned to love you a little in all those years of patient waiting?"

His hand steals up to that resting on his shoulder; leaning back to look in her face, he repeats the query in English—not Arabic—adding, merrily—"I believe it myself!"

The dark, piquant face glows red as a Georgian peach; the plump hand falls smartly upon his cheek; the black eyes are full of confusion and laughter:

"No! no!" she cries, unguardedly. "Never! never! not one bit!"

The hero of the idyl of the house-top flashes a triumphant glance at us:

"Didn't I tell you she understands English, and could speak it if she would?"

CHAPTER XXIV.

"THE WAILING PLACE."

THE deep descent down which we pick our way to the section of the ancient wall of the Temple known the world over as "The Wailing Place," is nothing more or less than an accumulation of rubbish hundreds of years old. Ancient Jerusalem lies from fifty to seventy feet below the modern town.

We turn aside from the more direct route to the sacred spot for a look at "Robinson's Arch." The celebrated fragment springs boldly from the wall which enclosed the area of the Temple. The bridge of which it was an abutment once spanned the valley between the House of God and Mount Zion. The polished stone pavement below is overwhelmed by nearly fifty feet of earth and stones. This bridge, in the opinion of able commentators, was but one of several arches connecting the two hills, probably the "ascent by which Solomon went up into the house of the Lord."

When the eager spirits that direct and execute the work of the Palestine Exploration Society so far prevail with the jealous Government now in despotic possession of the storied precincts as to be allowed to lay bare the bosom of the City of the Great King, what marvels will be brought to light!

In discussing the subject for the twentieth time in three days, we almost fancy that the conscious superincumbent soil thrills and heaves beneath us with a sense of the importance of buried secrets. We recount wistfully the timbers and boards of cedar of a day when silver was in Jerusalem as stones, and cedars as sycamore trees for abundance; the cherubim carved out of olive wood and overlaid with gold; the wrought cornices and wainscoats of cherubim and palm-trees and open flowers; the doors decorated in like manner and overlaid with fine gold beaten into the underlying tracery; the three rows of hewed stone; the brass work of flowers, of lilies, lions, oxen, cherubim and palm-trees, knops and pomegranates, the marble and ivory, the ceilings of cedar, and walls of vermilion. Nebuchadnezzar and his imitators in unholy demolition and robbery must have overlooked much in sacking Temple and palaces; the flames and Roman conquerors spared, because they could not destroy the wonderful stones of Herod's building.

Day and scene are in unison with this train of thought. The alleys we have traversed since leaving David Street, twist in and about the Jewish quarter. The last twist brings the Wailing Place into view.

A blind wall lifts high above our heads the great stones we have learned to recognize as belonging to Phœnician and Jewish architecture—some twenty, some

"THE WAILING PLACE."

thirty feet long. They were built so "compact together" that but a few crevices give foothold to hyssop and other clinging plants; they are darkened by rain and drip with moisture which we fancy is unclean. Close against the base of the wall is ranged a row of figures—men, women and even children. Women, for the most part past middle age, wear bedrabbled skirts, and faded shawls drawn cowl-like over their heads. Nearly all hold tattered copies of the Hebrew Psalms and the babble of intoned reading is broken to the attentive ear by sobs.

We know what they are saying, and, withdrawn a decent space from the throng, open the "International Teachers' Edition" of our "Palestine Baedeker," and follow the service silently.

O God! the heathen hath come into Thine inheritance:
Thy holy Temple have they defiled:
They have laid Jerusalem on heaps.
* * * * *
We are become a scorn to our neighbors;
A scorn and derision to them that are around about us.
How long, Lord, wilt Thou be angry forever?
Shall Thy jealousy burn like fire?
Pour out Thy wrath upon the heathen that have not known Thee,
And upon the kingdoms that have not called upon Thy name:
For they have devoured Jacob,
And laid waste his dwelling-place.
Oh, remember not against us former iniquities;
Let Thy tender mercies speedily prevent us;
For we are brought very low!

"Very low!" we interrupt ourselves to comment.

Among the dwellers in fallen Jerusalem, they are the most degraded, and to human eye, the most hopeless. The Psalm they chant might have been penned yesterday, so graphic is the portrayal of their present estate. For eighteen centuries the cry now upon their lips has been sobbed to heaven. Well may they iterate:

Wilt Thou be angry forever?

As our ears accustom themselves to the intoned clamor, we discover that three or four men stationed beyond one whose long robe and furred cap proclaim his priestly office, are conducting with him a responsive service quite irrespective of the readings going on about them. A learned friend in our little party translates the litany made familiar to him by many hearings:

For our temple that is destroyed,—

chants the leader. The response follows:—

Here sit we down lonely and weep.

WAILING PLACE AND UPPER COURSES OF STONES IN WALL

Then,

> For our walls which are torn down,
> (Here sit we down lonely and weep!)
> For the glory that has vanished,
> (Here sit we down lonely and weep!)
> For the goodly stones that are dust,
> (Here sit we down lonely and weep!)

A roughly laid stone wall closes the area at one end; a door set in it is slightly ajar, and withdrawing still further from the mourning crowd, we push it back and step into a sort of courtyard where are several trees. Beneath a branchy olive a woman sits flat upon the damp earth, her forehead laid against the trunk of the tree. A soiled and ragged book is upon her knees; her hands are clasped below her chin, and tears roll slowly down her withered cheeks upon the worn pages. Her lips move soundlessly in prayer, or penitential Psalm. She does not notice our intrusion, but a young girl standing behind her and looking up through tears at the wall, starts nervously aside, turning her face away in timidity or distress. A baby is in her arms; both are wretchedly clad and look half-starved. The guide knows her by sight.

IN THE JEWISH QUARTER.

"She is but thirteen years old," he tells us, as aware that we are trespassers, we withdraw, pulling the gate shut after us. "That is her grandmother, and, young as she is, the child is hers. Her husband left her before the baby was born. They are very poor and almost friendless."

"Are they Jews?"

"Of course; but not natives of Jerusalem. From Poland, I think."

Nothing in the strange scenes of the hour will linger more distinctly in our minds than this little episode. Compared with the unfeigned grief of the aged exile bemoaning her nation's fate in the one retired spot from which she could gaze upon the sacred stones, the loud lamentings of the line ranged against the wall, seem artificial and ostentatious.

"A periodical parade of mawkish sentimentality!" say strident accents at our elbow. "At the best, the whole business is the outcome of custom and superstition and unworthy of respect. Is it reasonable?—I go further—is it possible?—

to believe that these people come here every Friday ready to shed tears of genuine distress over a story almost two thousand years old? It is all done for effect—pumped up to order. If nobody ever came to look at the show, and travelers never wrote up 'The Wailing Place,' the exhibition would go out of fashion inside of five years."

Dr. L. Zwinglius Sharpe is, of course, the speaker. He is robed in a mackintosh which owes something of its exceeding glossiness to the Scotch mist thickening in the valley. His wife's gray cloak and hat have a conventual touch; her placid face might be that of a Dutch Madonna.

"We have seen *nothing* finer, more elevating than this, my love," she sighs blissfully. "Everything harmonizes *so* consistently! The hoary old walls, the bright colors of the women's shawls toned down by the wet—the solemn sound of the chanting—and oh, *Doctor* dear! *will* you look at the *delicious* dull olive of those old velveteen coats? And that woman, over there sitting on the stones—the one in white, who is looking this way—*doesn't* she remind you of 'By the rivers of Babylon we sat down?'"

From the rising ground about a stone's cast from the "Wailing Place," we glance back. The crowd is thinning, and we cannot avoid seeing that some of those who mourned most volubly are most brisk or indifferent by the time they reach us. There is no difference in the aspect of those who were but now apparently plunged into the depths of hopeless sorrow and others of the race who are buying and selling at the stalls in the Jewish quarter.

"Is Dr. Sharpe right for once?"

While we say it, the grandmother and the child-wife we surprised at their devotions, come slowly up the narrow way. The baby has fallen asleep upon the mother's shoulder. The load is heavy for the little creature, and stopping to get her breath half way up the hill, she turns for a last look at the grim wall below. Tears well afresh into her eyes, and again she averts her face, as if ashamed of the weakness. The old woman has drawn her dingy shawl over her book; trudging weakly onward, she keeps her regards fixed upon the ground; the sadness of a long-frustrated desire is in the drooping lines of the wrinkled face. She is not artistically "harmonious," but we believe her to be the type of a class. Her figure and visage come between us and the page upon which we read, when we shed our damp wraps and sit down before a glowing fire in our hotel quarters:

> The Lord was as our enemy:
> He hath swallowed up Israel:
> He hath swallowed up all her palaces:
> He hath destroyed his stronghold.
> And hath increased in the daughter of Judah mourning and lamentation.

CHAPTER XXV.

A ROUND OF VISITS.

HAVING stipulated that our visits on this afternoon shall be confined to respectable people whose worldly circumstances are moderately good, I demur when Mrs. Jamal, holding up the hem of her clean white izzar, leaves a narrow dirty street for a cross-alley, narrower and dirtier yet.

When I say that a Syrian street is dirty, I seek to convey a depth of meaning hard to be received into the imagination of the dweller in Western cities. This particular alley is positively noisome. Refuse of the most objectionable description rots and reeks in corners and close to the walls. The least unsafe footing is in the middle of the street. The houses stand close together, are from two to three stories high, of stone, and badly lighted. Most of them have but a single window in each room; some have none, the door admitting all the light and air that find their way to the murky interior.

All the children who can walk alone are in the street. In commenting upon the circumstance, I check the censure upon my lips. Ailie Dinmont retorted, when her husband complained to a visitor that she "would aye give the bairnies their ain way," "It's little else I have to give them—puir things!" And the uncombed, unwashed, black-eyed elves, strolling and scampering in what is no better than a filthy gutter, would seem to have preciously little except their own wild will to make life endurable.

Six or eight troop at our heels as we mount to the third story of a building no better and no worse than its neighbors.

"I will take you first to see the people who own the house—people who are well-off, almost rich, for this part of the world," my conductress flings back to me over her shoulder, her pleasant eyes sparkling with fun.

I see the meaning of her arch look the next minute. We have climbed the stairway, built, as are all in the neighborhood, upon the outside of the house, and slippery with mud, and pause at an open door while my guide asks leave to enter. Two-thirds of the one-room constituting the "home" is raised by two steps from the rest, the door opening at the side of the lower portion. Near the edge of the upper step a woman is slicing carrots into a big metal bowl—copper or brass, but black as iron. Her legs are doubled under her; the floor is covered with dirty matting, and she sits flat upon it. She wears a dingy cotton skirt and a wadded jacket; upon her head is an equally dingy mendeel. A brazier of charcoal is not far off, over which the meal she is preparing will presently be cooked; a divan, cushioned with Turkey red cotton, runs around two sides of the apartment; in a recess is a pile of quilts and rugs for bedding. Under the divan are a water jar and an oil-jug, with three earthenware bowls of various sizes, also a coffee-pot and half-a-dozen cups. There is not another stick of furniture; not another utensil is to be seen.

"A STIRRING AND NOTABLE HOUSEWIFE."

Without rising, the hostess says in Arabic that we are welcome, and motions us to seat ourselves upon the divan, going on with her work, placidly ignorant of any deficiencies in her garb or surroundings. The children crowd together at the bottom of the room, staring stolidly at us; an older woman than the mistress of the establishment seats herself upon the matting, reaches out an arm to grab the solitary garment of an eighteen-month-old creeping dangerously near the steps, and when he screams and slaps her, smiles listlessly. From her appearance I should take her to be sixty years of age, and am amazed at discovering that the baby-boy is hers and not yet weaned. She was the sixth wife—and is now the widow—of a Moslem who could afford to please himself in the matter of a plurality of partners. Three were divorced, and two died while he was alive. Three of the children we now see belong to the widow by right of birth; her step-sons and daughters are married. Her jointure as her lord's only legal relict enables her to live independently of them, I am informed. The deceased was in easy circumstances.

The widow's idea of independent ease is to dwell as a boarder with the gain-loving proprietors of the three-story tenement. She and her children, the man of the house, his wife and two children, eat and sleep in this room. Upon hearing it I gasped. The place may be fifteen feet square. It is certainly not larger than this computation would make it. There is but one window, and that is set high up in the thick stone walls. A pair of pet pigeons are tip-toeing about the door, pecking at crumbs; there is even a dog upon the threshold, albeit he is an accursed animal among the Moslems. At night-fall the door will close all these creatures in, and the window remain shut for fear of the damp draughts. The floor will be covered with quilts, and eight human beings lie down to sleep in the clothes they have worn by day.

"Yet you tell me these are not of the poorest class? that hundreds of others live in the same way without losing respectability?" I inquire.

My interpreter raises her eye-brows, and makes a gesture that would be impatient were it not courteous.

"I do tell you that they consider themselves comfortable. They have houses and money. They want nothing. And are proud—so very proud that did they know that we are here to see how they live, we would not be let inside of that door. I have told them that you are my English teacher—" and at my look of surprise, she adds, gracefully, "Indeed, madam! it is I who have learned much—ah, so much from you! I know these people. They have good characters and kind hearts. They are not of my religion, but I respect them. You wished to see the ways of the country, and you are seeing them."

This is uttered low and rapidly, with the impassive countenance we train ourselves to assume when upon such errands. But for the speaker's womanly wit and tact, I could not gain access to what I especially wished to inspect, the homes of the great middle class of Syrians. I bow to the unintentional rebuke conveyed in her protest. If "these people" choose to sit cross-legged and eat and sleep upon the floor, it is surely their affair and not mine. My prejudices in favor of carpets, mattresses, tables and table-cloths, knives, forks and bath-tubs, would seem as outlandish to them as their peculiarities appear to me, while as to their "pride"—the wife of an American freeholder would resent yet more indignantly the intrusion upon her domestic life of a Jerusalem reporter—if there be such a thing.

I copy carefully Mrs. Jamal's deportment, as she takes leave of our entertainers, after ten minutes' chat in Arabic. I even utter after her the one Arabic phrase I have learned, "*Mar Salaami!*" in withdrawing our feet from the door.

Visit No. 2 has livelier elements in it. In point of worldly circumstance the families represented in the group to which I am introduced are apparently superior to the people I set down as "No. 1." The room, approached, as before, by an outer stone stairway, is larger than that we have just left, but looks smaller for having an iron bedstead, draped with mosquito netting, and a veritable bureau, in it. A plump cushion, covered with Turkey-red, is laid upon the floor across two sides of the room, clôse to the wall. Upon it sit, cross-legged, the grandmother of the household, sewing upon a man's shirt, and a Nubian woman, as black as charcoal. A deep brand upon the left cheek shows that she was once a slave. Although manumitted by law some years ago, many African slaves have elected to remain with their former owners. This is evidently one of the family. While her mistress plies the needle, the negress folds her arms upon her stomach and leans lazily against the wall, a grin upon her face, and more curiosity in her rolling eye-balls than I have ever seen evinced by a native Syrian of either sex. Chairs are set for us by a young woman who enters hastily behind us. She is introduced as a married daughter of the old woman. Her room is upon another floor of the house, and this is virtually but one family.

WOMAN WAITING WITH BOWL OF DOUGH AT A CITY OVEN.

The invariable brazier is in the middle of the matted floor, and the unavoidable coffee steams upon it. We have talked of the weather and the chances of more rain for five minutes before the coffee is poured into tiny, handleless cups by a representative of the third generation, a girl of fourteen who is to be married in a couple of weeks. She trips around the circle in stockinged feet, having left her sandals at the door; her calico gown, belted about a full waist, is clean, and very much like what a New England farmer's daughter would wear while doing the morning's housework. Her brown wrists are banded with gold and silver bangles, eight or ten upon each arm; she looks really animated as, in taking the cup from her hand, I wish for her, through our interpreter, many years of happiness in her new life. The spirit of enterprise exemplified in bed, bureau and chairs, gives tone to the household. A breeze from the world outside of harem windows has stolen in here through some unguarded crevice. The talk has briskness and almost sprightliness in it. The coffee is thick after the manner of all Turkish coffee; it is, moreover, made excessively sweet out of compliment to the foreign guest, and a liberal pinch of allspice is added to it. A decoction of wormwood would be more palatable to one of the victims, but the alternative being to take one of the cigarettes assuming form in the supple fingers of the daughter, the draught is committed to an amiable, and just now oft-abused, stomach.

The cigarette-maker has deposited her yearling baby upon the floor, and, with a bowl of tobacco upon her lap and a pile of papers upon one knee, wets forefinger and thumb in her mouth before beginning each fresh roll. Her mother lays by her sewing; the Nubian loosens her lazy bones to accept a cigarette; the bride-expectant puffs as fast as the rest, and the baby does not cough when the room is soon blue with smoke. The top of the bureau is ornamented with three tripods of black wood inlaid with mother-of-pearl, all standing upside down. One is placed in front of me, and several cigarettes and a box of matches are laid in seductive array upon it. Should I change my mind, overcome by the temptation, I need not be mortified by the necessity of asking for the luxury.

The cigarette is always offered to me, and, I need not say, never accepted. My reputation for taste and good manners must suffer in the minds of the spectators, but they refrain from criticism, spoken or looked.

Coffee is made every hour in this room, and cigarettes are rolled by one or other of the women all day long. They do literally nothing between meals, but sip and smoke and—as a pasha's wife told a visitor of her own method of passing the time—"just sit." The grandmother is a notable and stirring house-wife in that she knows how to sew, and occasionally makes a garment for her soldier-son. She is further distinguished among her neighbors by preferring to sleep in a bed rather than on a quilt upon the floor.

The greeting bestowed upon Mrs. Jamal is cordial to affectionateness. She explains it aside, presently, the baby's screaming demand to be taken up and fed absorbing the attention of the others.

"You see I run in once in a while, and tell them stories. They are only children—nothing more—and will listen for hours to tales of Abraham, Isaac, Jacob, Joseph, Samuel, and all that, you know. I give them only Bible stories, but that they never suspect. Not that they would care, but the men object to having their wives and daughters taught the Christian religion. They say it makes them too independent. So, I must slip in the good I would do them—poor souls!"

A quick footstep sounds upon the stone stair. The women of three generations arise to greet the stalwart young fellow in uniform who appears in the doorway. One after another, they kiss his hands; his aged mother hastens to set the one unoccupied chair for him; the young niece serves him with coffee, his sister supplies him with a cigarette and lights it for him. He is not ill-looking, and his behavior is "quite the thing," according to his lights and those of his kinspeople. He has nodded to us, receiving without a word the attentions paid to him by his assiduous servitors, and when we exchange "*Mar Salaamis*" with the rest of the party and are urged to repeat the call, we leave him seated, his long legs stretched

COURTYARD OF A HOME IN NAZARETH.

half-way across the room, the cigarette between his lips, his handsome eyes fixed upon the ceiling.

No. 3 of this one of our "afternoons out" is paid to a young matron who would be comely were she less slatternly and plumper. Her room is without a window, and if the sun did not shine out brightly for the quarter-hour of our stay, we should not be able to see from one wall to another of the gloomy confines.

My guide knits her brows disapprovingly and sighs in entering. Her second remark to the hostess is a protest. A cradle is drawn near the door that the light may fall upon a weazen-faced baby, sitting upright in it. His eyes are startlingly large and dark in the small visage; the fingers eagerly outstretched to the bowl in the mother's hand, are like the claws of a bird. He is too hungry to notice the strangers, yet Mrs. Jamal intercepts the vessel.

"You are not going to give him this!"

She shows it to me.

"Curdled!" I say. "We call it 'loppered milk!'"

"It is lebben, and old lebben at that, sour like vinegar and mouldy, and not too clean. Child! child!" turning to the mother, "don't you know that babies should have sweet milk? the freshest you can get? Goat's milk, if you cannot get cow's?" And again to me and despairingly—"What can be expected of a mother who is hardly more than a baby herself? She is but fourteen now."

The girl-wife laughs childishly and good-humoredly, and proceeds to hold the nauseous-looking mess to the greedy lips.

"He eats it every day. He eats nothing else," she says carelessly.

"He looks like it!" retorts the visitor. "Ah, madame, it is cases such as this that make the heart ache. I have known this girl since she was six years old. She was married at twelve. She cannot cook, she cannot sew; her baby's clothes must be made by somebody else, and you see what a place this is for a husband to come to at night. I would teach her if I could, but they never learn anything after they are married."

The mother looks around at the earnest tone, and laughs again, comprehending nothing except that her friend is troubled, and about her, and profoundly indifferent to censure as to praise. The chamber is as ill-furnished as her mind. No "Franji" innovations, in the shape of beds with legs to support them, seats to match, and chests of drawers, have perverted Syrian simplicity here. A roll of dirty bedding in one corner, a brazier of hot ashes, the coffee-pot simmering upon it, a big bowl, the cushioned bench against the wall, and the cradle, make up the list of household goods and chattels. The odor of unwashen humanity; of past cookery, in which onions took an active part, and of stale tobacco-smoke, blend queerly and powerfully with the pungent fumes of the boiling coffee. Seasoned as I have thought myself to be by now to assaults upon the olfactory organs, I am thankful to be allowed to retreat to the outer air. My companion lingers a moment longer to explain that we have not time to accept the coffee the good-natured hostess would press upon us, and to offer a few more words of neighborly counsel.

"Not that it will do any good," continues the mentor in rejoining me. "But I cannot help liking the poor child. She is of such a good disposition and loves her baby. Her husband is much older than she and has steady work, and would be kind to a woman who made his home comfortable. I try to make her understand this, and to encourage her to keep things clean and learn to sew and cook his food well. As you have seen, she but laughs. She ran wild upon the street until she was married, just as those girls are doing now"—directing my notice to a knot of children scrambling over a broken wall at the bottom of the street, chattering and shrieking like so many parrots. "What can mothers expect who let their daughters grow up so, but that they will have homes and babies like those you saw just now?"

She is so much absorbed in the subject, and so intent upon shielding her white izzar from the pollution of pavement and wall, that she does not at once hear a soft call from the landing of a flight of steps bulging over the street.

"SHE RAN WILD UPON THE STREET."

The summons is repeated, and we look up to see a woman beckoning importunately. She pulls her mendeel forward over her face and runs down to meet us, when we begin the ascent of the stairway.

"She will have it that we must come to see her," smiles the interpreter, introducing me in dumb show.

My new acquaintance unveils her face when we are within her door, kisses my hand and Mrs. Jamal's cheeks, and begs us to be seated upon the divan. It is covered, I note immediately, first with the conventional Turkey-red, and over this with clean, white muslin, a sort of "scrim," such as is used for window-curtains. An iron bedstead fills up one corner of the room, curtained with the same material as the divan-cover; a neat chest of drawers is in the opposite corner, and a tray set upon this holds cups, saucers, tumblers and plates. A cupboard is let into the wall; a baby is creeping over the floor. Mrs. Jamal picks it up and kisses a face that is positively clean and almost chubby.

"The youngest of four!" she interjects, between caresses and pet-names. "The mother was married at ten. Her first was born at fourteen. She is nearly twenty now. From eight until ten, she was in my school. She reads and is handy with the needle. And we have always been dear friends. Ah!" nodding and smiling to the pleased mother—"I am very proud of my scholar!"

We have been seen from other doorways. Mrs. Jamal still cuddles the baby when two more women enter; one leads a child; her companion carries another in her arms. Salutations and introductions over, the mother of the toddler prefers a request. A piece of dark blue cloth is thrown over her shoulder, just purchased for her by her husband, and she would like to make a jacket out of it for herself. With frank disregard of conventionality, or what we account conventional usages, Mrs. Jamal lays by her izzar, spreads the cloth upon the matted floor, kneels down beside it, measures and chalks a pattern upon it with workmanlike skill, and cuts out the garment as swiftly and well as if she were a trained tailoress. The women follow every movement with keen interest, and I am almost as much engrossed.

It is doubtful if my conductress has ever heard of "zenana missions," under that name. She would be surprised were I to dub her "a home-and-foreign-missionary." She is merely a Christian woman, dwelling among her own people; a housekeeper, wife and mother, happy in her home, and longing to make other homes as happy as hers.

I have already learned to distinguish Christian women from Moslem wherever I see them, and silently conclude (correctly, as I learn presently) that the bearer of the baby belongs to the Greek Church. It is not an expression of superior intelligence alone that characterizes the Christians, but a general alertness of bearing and play of countenance, a sort of self-respectfulness, if I may thus phrase it. "This woman has a soul of her very own," it seems to say. "She thinks, feels, hopes and lives her own life to some extent. Knowing herself to be immortal and more accountable to God than to her husband for word and action, she is lifted above the beasts that perish."

It is rare to find a Moslem woman of twenty who can read or write. It is rarer to meet with a Christian woman of any age among the respectable middle classes who cannot. A teacher in a mission school told me yesterday that a wealthy Moslem, yielding to his daughters' importunities to be sent to school—education of some kind being just now a "fashionable fad" in their circle—brought them to the English lady in person, and asked that they might be instructed in needlework and music and even taught to read.

"But not to write!" he stipulated. "Women who can write will bring disgrace upon their families by sending love-letters to men who are not chosen by their parents to be their husbands. It is never safe."

Let me remark, furthermore, while my guide kneels on the floor, the shining shears moving as briskly as her kindly tongue—that jealous dread of the encroachments of the Christian religion is on the increase, not decline, among Mohammedan men. Mrs. Jamal conveyed the secret of this distrust in saying that they fear religion will make wives and daughters independent. Dr. Sandreczky, whose Children's Hospital is one of the noblest monuments to Christian enterprise in Jerusalem, and to whose courtesy and professional skill foreign residents and visitors alike bear grateful testimony, once said the same thing in effect to me:

HOLIDAY STREET SCENE IN JERUSALEM.

"In the Moslem's opinion, the most mischievous idea that could get possession of the mind of his wife or daughter, would be belief in the equality of sexes. Whatever tends to overthrow the tenet of masculine—that is, patriarchal—sovereignty, is a direct blow at the Mohammedan religion. The woman who suspects her own individuality is dangerous."

Here is the key-note to what makes mission work among the followers of the Prophet especially difficult and discouraging. It is not uncommon for an English or American visitor to a harem, small or large, to be warned not to speak of Christ or his religion to the women, and frequent calls from the pale-faces would excite a degree of suspicion which would lead to exclusion. It follows, as a necessity, that the good done to the creatures who minister to the needs and pleasures of the superior sex, must be wrought tactfully and gradually.

THE DAMASCUS GATE, JERUSALEM.

"I am no more in my husband's sight than his donkey!" said a sad-eyed woman to an American woman who had been an angel of mercy to the miserable household. "Not so much, for he beats the donkey less."

In our homeward walk, I say something of this to my companion. She shakes her head mournfully:

"You see but a little. If you were to live among them year after year, you would be very sad. I do what I can to lighten the load of my small world. But what is that among so many?"

Three times a week she collects in her neat sitting-room from fourteen to eighteen native girls, and teaches them to do plain sewing, embroidery, knitting and mending, giving her house and her time without charge. The girls take turns in cooking a luncheon served at noon, washing up the dishes and setting things in order when the meal is over. In the spring, she leads them into the country to gather the wild flowers that glorify the hills and fields for a few weeks. These are carefully pressed, and, under her eye, arranged upon cards inscribed, "Flowers from the Holy Land," and sold during the tourists' season as souvenirs of the visit. The sales of these and other articles do not meet the expenses of materials and food, unless the number of visitors be unusually large. In an ordinary season, the class is kept up at a loss. What afflicts the instructress most is that she must refuse many applications from girls who wish to become pupils, would-be recruits from the ranks of the empty-handed and empty-minded, whom the busy philanthropist pities from the depths of her generous heart.

Mrs. Jamal's modesty would shrink from the title of philanthropist. She is simply living her life conscientiously, in the fear of God, and in love for her neighbor in the sphere to which she has been called. Yet I can answer for her gratitude should the proposition made by me in a former "Talk"—that some of the "Circles" in the far land in which women are surely more highly favored by Heaven than in any other, would send monthly a few dollars for this quiet, practical work—be taken up in active earnest. Twenty-five dollars per month would enable the teacher to enlarge her class and increase the usefulness of what she has undertaken.

She can labor successfully where foreigners may not enter, and, being one of themselves, can brighten and better her country-women's lot as it now is, while gradually raising them to desire and appreciate higher things.

CHAPTER XXVI.

IN A PALANKEEN TO JERICHO.

THE romantic reader of Oriental fable and travel conjures up at the word "Palankeen," a vision of luxury which will be dashed by a description of the conveyance as it is brought up to the hotel door on this fine frosty December morning. It is not a curtained litter, piled with silken cushions, borne upon men's shoulders; neither is it a sedan chair swung between gilded poles. A stout wooden box, open in front, and half-way to the back on the sides, is fitted upon two pairs of shafts, one behind, one in front. A glass window, working upon hinges, is let into each side, and over these hang red curtains. A bench fastened into the rear end of the interior is cushioned with red stuff; a mat lies on the foot.

At first sight the vehicle strikes one as clumsy, unwieldly and grotesquely ugly. A mule is harnessed within each pair of shafts; rugs, a railway pillow and my satchel are bestowed within, and I mount to my carriage by means of a step-ladder, held firm by David himself. I was inclined to laugh when this feature of Oriental magnificence was first submitted for my approval, but in previous journeys to that upon which we are now bound, I have learned to appreciate the shelter and comfort it affords, and respite from the saddle when one is weary, or not well, is grateful in the extreme. Packed in among downy cushions, swathed in furs and soft woolens, with abundance of room for limbs that tire of one position in the long marches, I do not begrudge Mrs. Lofty her carriage and high prancing pair, as the muleteer gives the word and the little caravan sets forward through the Jaffa gate. Some women have complained to me that the swing of a palankeen made them slightly sea-sick for a few hours. Although a wretched sailor upon the water, I have never known the slightest nausea in this ungainly equipage. The motion is precisely what one feels in the saddle on the back of a fairly well-gaited horse. It is not as smooth as that of a carriage built over elastic springs, but the comparative roughness has ample compensation in the roominess of the interior and opportunity for change of position when one is cramped by sitting.

We were once jocosely critical over the quantity of timber in the body and shafts of my "coach." Experience has proved the need of stanchness and solidity; nevertheless, as we jog along the first familiar stages of our route, I fall to musing upon what manner of conveyance I would have constructed in America, and presented to my faithful dragoman, had I the means to compass this end. It should be light and strong, weather-tight, and upheld by such springs as only American carriage builders know how to make, and be gratefully dedicated to the use of American women tourists.

A halt is called, and I arouse myself to inquire why. David, gorgeous in professional gear, rides to my side to prefer the petition of a native photographer to "take" the palankeen and the escort. The picture herewith presented gives so correct an impression of the procession and the place—a stone-strewn olive-orchard on the lowest slope of Olivet—that I commend a study of the novel scene to the curious and interested. The camp and camp appointments have gone forward, leaving to me as attendants, Serkeese, grinning from the top of the swathed luncheon-tent and hampers; the two muleteers; Alcides, upon Massoud, and disguised in hunting shirt, fez and kafeyah, and David Jamal upon Dervish. The foremost posts of the palankeen are wreathed with leaves and flowers in honor of our "fresh start."

THE LEPERS' HUSKY CRY.

"Now," calls out Alcides from the van, as we are released from the glaring eye of the camera, "We can all go to—Jericho !"

"Quite so, sir !" replies the only hearer besides myself who comprehends an English sentence, a grave response that may or may not signify familiarity with American slang.

16

"OUR FRESH START."

For our destination is Jericho, and we are to be absent from our headquarters in Jerusalem for some days. We pass Gethsemane, dispensing a few coins to still the husky shrillness of the lepers' cry; skirt the Mount of Olives, and, diverging to the right, fall into line upon the Bethany road, an excellent thoroughfare for several miles—almost half-way to Jericho. After we have left the mean hamlet of Bethany behind us, we begin the descent between hills that grew more and more barren, as we go on. The olive is the only tree to be seen, and the groves of this become rarer and smaller. At the juncture of the Bethany highway with a comparatively disused track leading to Bethphage, waits a mounted and dignified figure—the sheik who is to be our safe conduct as we go "down from Jerusalem to Jericho." He and David ride together; Alcides falls back to the side of my palankeen; Serkeese trots on ahead with his queer looking load. It is a taciturn company when we are bound upon an all-day expedition. The road winds down into a tortuous valley, at the head of which David reins up his horse and waits our approach. A small, half-ruined khan planted upon the edge of the way has a reason for being in a well opposite to it, and lower by twenty feet or thereabouts. Both are very old, the situation of a khan usually remaining the same for countless generations, particularly when it is in the vicinity of a spring of living water.

SHEPHERD AND WIFE.

"THE FOUNTAIN OF THE APOSTLES."

(244)

Some one at the hotel spoke of this well last night as probably En-shemesh, one of the division lines between Judah and Benjamin.

"The waters of En-shemesh, and the goings out thereof, were as Enrogel." ("The Fuller's Fountain," reads the margin.) "And the border went up by the valley of the son of Hinnom unto the south side of the Jebusite; the same is Jerusalem."

Our guide has a more interesting story of the waters:

"Called 'The Apostles' Fountain,' because of the tradition that Christ our Lord, on His last journey to Jerusalem, rested here while telling His disciples that

RUINS OF ROMAN WATCH-TOWER.

He was going up to be betrayed and crucified. He was on His way from Jericho, you see, and not yet come to Bethphage, near the Mount of Olives. It is only a tradition—but it might have been."

"It might have been," for the ruined wall and part of a roof of the ancient caravanserai are distinctly Roman, and the spring is perennial. What more natural than that the "taking of the twelve disciples apart in the way" may have been at the foot of the ascent upon the summit of which He was to meet the "great multitudes" and their shouts of "Hosanna to the Son of David! Blessed is He that cometh in the name of the Lord!"

Be this as it may, this was the route by which He came up from Jericho where He had dined with Zaccheus, and in departing from which He had made the blind Bartimeus to see.

All signs of human habitation fail us as we proceed. High hills line the highway, ridged by natural terraces, treeless, and at this time of the year, sere to the summits. The heights, with their naked ledges, remind us of nothing more forcibly than of gigantic skeletons from which the flesh has wasted and dropped away. The silence is profound and awful. The clink of the horses' iron hoofs upon the rock-bedded road and the harsh call of the muleteers to the patient beasts are all that we hear for miles together. Another—and a credible—tradition is committed to us in a defile, an hour's journey beyond the Apostles' Fountain. Hereabouts, it is believed, Shimei came out from Bahurim, "and as David and his men went by the way, Shimei went along on the hill's side over against him, and cursed as he went, and threw stones at him and cast dust."

As we read the tale, imagination is quick with the picture of the royal father's flight between the hills, and there is cordial, although may be, unholy sympathy expressed with rude Abishai's outbreak: "Why should this dead dog curse my lord the king? Let me go over, I pray thee, and take off his head!" We can almost see the foul-tongued son of Gera—whose behavior, after the king's restoration was a convincing proof that the bully is always the coward,—leaping from one to another projection of the hillside, to hurl down missiles upon the fugitives who "came weary and refreshed themselves there."

On, on we ride, between sad and silent heights, over waterless ravines, whose pebbly beds will be washed and tossed by turbulent floods after the spring rains. The glare of the white rocks is now and then relieved by curious pink strata, sometimes deepening into rose-color. At long intervals, we see high up on the breast of the mountain mingled flocks of sheep and goats, feeding together, while the shepherd, wrapped in his brown-and-white mantle, stands or sits near. Here, we observe for the first time, an illustration of the division of the mixed herd prophesied by our Lord, and to which our attention was once called by Dr. Webster, the devoted missionary of Haifa. A shepherd, walking before the flocks, a sickly lamb tucked away close to his warm body inside of his thick garment, called them, as he went, in a peculiar note, plaintive, yet penetrating, which we could hear plainly half a mile away. The sheep followed him in a long steady line,—the black goats straggled hither and yon, cropping the herbage, running races, scampering up rocks at the left and right, until we wonder how he could ever collect and finally get them safely to the fold.

"And when he putteth forth his own sheep, he goeth before them, and the sheep follow him; for they know his voice."

Alcides spurs up a hillside to have audience with a shepherd, whose garment of goat's hair is wrapped about him as gracefully as ever Roman senator wore toga, and gallops back in proud possession of a veritable pastoral pipe. The sounds evoked from it by the new owner are weak and shrill, yet "carry" well in the still air.

"Do you suppose that the boy-shepherd of Bethlehem played psalm-tunes upon a machine like this?" winds up the performance. "One could get better music out of a penny whistle."

The classic instrument is about eight inches in length and consists of two reeds, each pierced on one side with six breathing holes. They are fastened longitudinally together with resin, and into the upper end are fitted two movable mouth-pieces, attached to the pipe by a twine string. It is a primitive affair, but looks as if it might have capabilities—in the right hands.

"THE SHEIK MOUNTS GUARD WITHOUT."

Noon is near when we espy the Hathrur Khan, or the Khan of the Good Samaritan. Having always supposed the story of the man who fell among thieves upon this road to be merely a parable related with a specific purpose, we are surprised to learn that it is here considered that our Lord made use of a real incident, as historic as the fall of the Tower of Siloam, in answering the lawyer's question—"Who is my neighbor?" The Jericho road has been for twenty centuries the haunt of banditti. Even yet, peaceable husbandmen conveying crops to the Jerusalem market, or returning with the proceeds of their merchandise, are often set upon by watchful predatory tribes, and robbed. It is certain that the site of this wayside khan is very old, and that it has borne for hundreds of years the name of "him who showed mercy" to a feudal foe. It is almost certain that our Lord, journeying, as usual on foot with His disciples, stayed His progress to the city, where He was to offer the last and Greatest Sacrifice for sinning humanity, and rested here, perhaps for the mid-day meal, as we are now doing.

The khan is a big shed, with a courtyard at the back, as new as it is uncomely, but the foundations are ancient. Upon an overlooking hill are the ruins of a Roman watch-tower from which is an extensive view of the defiles of approach in all directions.

Palankeen and mules are left without; human beings lunch in the vast, bare chamber, open at the back toward the courtyard in which the horses are tethered. David hovers about us to be sure that we are comfortable; the sheik mounts guard without, perched in a niche of the archway, cigarette between his fingers, his eyes closed in meditation or slumber. The hum of conversation from another party of tourists and the chatter of the muleteers in the road, disperse thinly into the vast silences that encompass us in what is, for all we can see or hear, desolation without an inhabitant.

CHAPTER XXVII.

STILL ON THE JERICHO ROAD.

BEYOND the Khan Hathrur, the road from Jerusalem to Jericho grows more difficult until narrowing, almost without warning, into a bridle-path, it ends the speculations in which we have indulged as to the possibility of going by carriage all the way to the Jordan. So rough and straight is it that more than once I insist upon alighting and walking around a sharp spur where the track is a narrow ledge overhanging a precipice, or the hill we are to descend seems, to my apprehension, to have more incline than the side of a house.

The brooding silence is oppressive to ear and spirit. Alcides avers gloomily that the "best Baedeker for Palestine" opens of itself at Jeremiah whenever we would consult it. Even a cluster of black tents would be "comely" to eyes wearied by the illimitable stretch of treeless mountains, but we see none. Following the example of their robber ancestry, the Bedouins burrow in ravine and "wady," and natural dens and caves of the earth. The shepherds far up the heights, a few men on foot and still fewer upon donkeys are all the fellow-creatures we see until, in the dreariest part of the route, a defile so deep that the sun never visits it in winter except at noon, a woman rides briskly past. Being portly of figure, she is no inconsiderable load for her donkey, and parcels of divers patterns and weights are strapped to her saddle, but she goes by our party at a smart amble and is soon lost behind the bulging shoulder of a hill. Two men, similarly mounted, and a pack-donkey hardly visible under his load of baskets and sacks, follow her.

"An enterprising lady," the guide says, admiringly. "She and her brother keep the hotel at Jericho. And keep it well. She has been to Jerusalem for supplies."

The incident is incongruous with the stern desolation of the scene, but we are glad of the relief of a laugh. A modern hotel, household supplies of sugar, eggs, flour, and coffee, not to dwell upon the brisk donkey and his plump rider—bring us down--(or up) from meditations among the tombs of a mighty Past. The cheerful ring of our voices comes back in a shocked echo from the furrowed walls closing us in on either hand. The upper heights are honeycombed with caves, winding paths, slight threads at a distance, leading to black mouths of some. These were, ages ago, the cells of hermits, emulous of austere Elijah's fame, and little dreaming, amid fastings and prayers and temptations, of the far different warfare to be waged against flesh and blood by their successors in the tenantry of the high places of the rocks.

Down the Wady Kelt which, by and by, runs parallel with our road, wears a fierce little brook, brawling so noisily that we can hear it before we can see it.

"The Brook Cherith," David names it, and we have enough to think and talk of for the next mile. We alight then, and quitting the track, clamber over intervening hills and gulleys to reach the cliff at whose base foams the fussy torrent. It will shrink to a tiny rivulet in summer, and is making the most of its day of power. Right over against us, scarcely distinguishable in the shadow of the perpendicular wall opposite from the rock to which it clings, is a miniature edition of a Greek convent. Those who founded it believed, or said that they believed, the site to be near Elijah's hiding-place during the miraculous drouth that plagued the land in Ahab's reign. They named it, however, the Monastery of St. John, the Elijah of the New Testament. It is marvelous how the materials for building even so tiny an edifice were transported to the chosen spot, although we trace again upon the face of the precipice thread-like paths along which wild goats might fear to tread. The monastery seems actually to "back up" against the grim wall as if afraid of falling.

Pink and rose-color have given place to chalky whiteness in the towering sides and heads of the mountains hemming us in. We are going down all the while, following generally the capricious turns of the brook, and never out of sight of the gorge it has worn in the everlasting hills. I have alighted to walk down a rocky fall—I cannot call it a descent—the abruptness of which makes me dizzy when I behold it from the palankeen; David Jamal and the sheik have gone on down, their horses treading warily upon the rolling stones, when half-way to the bottom we take a sharp turn and another world opens upon us. It is as if the two peaked promontories in which the sides of the pass terminate were rolled back like gates, and the landscape unfolded panorama-wise before our eyes. The plain of the Jordan loses itself behind sand-hills on our left, and farther away on the right is the Dead Sea, blue and serene as the Mediterranean on this afternoon.

Glimpses of the sacred river sparkle between clumps of green. The oasis, luxuriant upon the banks, disperses in sparser verdure and coppices of thorn trees receding from the water. The white tents of our camp gleam near one of these groves, and to reach it we must cross a plain overgrown with bushes, nearly a mile in width. The track is sandy; the mules with the best intentions with regard to their waiting supper, cannot make much speed. I am ashamed of the folly, but I have not been so undeniably terrified since first setting foot upon Syrian soil, as in discovering that the moving brown objects partially concealed by the low growth are camels—a herd of them—pasturing with nobody near to hinder any demonstration they may take it into their hideous heads to make toward the equipage which excites their curiosity. Utterly ignorant as to the stampeding proclivities of my favorite abominations, I do not know whether to expect a charge in force, or a challenge to single combat. There are, I judge, at least one hundred of the hulking creatures munching the coarse grass, stripping leaves from the bushes and principally—or so it appears to me—wandering aimlessly over the meadow in quest of some stray bit of mischief peculiarly suited to the camel-ine nature.

"A MINIATURE EDITION OF A CONVENT."

All three of my mounted escorts have ridden on, Jamal, doubtless to make all ready for our reception in camp, and the sheik for Jamal's good company. With difficulty I restrain the wild impulse to scream to Alcides' disappearing figure as Massoud gallops after the pair now lost to sight in the coppice. Glancing through the window on my right hand, I see a particularly malevolent camel, a big blackish brute, "point" me fifty feet off, and come straight toward the palankeen, nose outstretched and shoulders hunched up after the manner of that most idiotic of feathered bipeds—a turkey-hen. In a truce a dozen other noses take the same line, and a row of the horrors is coming directly for us. The muleteers pay no attention to them, until my energetic gesticulation directs their eyes to the peril. Then both laugh a little, and one shouts carelessly at the foremost camel, who halts to stare at us—and with joy unspeakable, yet which I dare not confess, I behold the top of Alcides' kafeyah above the bushes that swallowed him up, just now, and he canters back to see what has become of me. With all the consideration for my welfare that never sleeps or wavers, he has not bethought himself of my antipathy for the desert craft.

"What harm could they do to you? They are the most peaceable of animals," he chides, laughingly.

I offer no excuse for the silly terrors, from the effects of which I do not fully recover until the cup of four-o clock tea Imbarak has ready when we dismount at the tent-door, has been swallowed and the exceeding beauty of the view begins to act soothingly upon jarred nerves. The absurd incident has peaceable fruits of righteousness in one particular, at least—I shall never again sneer at a woman who jumps upon a chair at sight of a mouse, or screams when a spider runs over her neck or cheek.

The situation of the camp is, as usual, well chosen. We are, as I have indicated, upon the outskirts of the oasis. Beyond the band of verdure, apparently so near that we are incredulous as to the miles that separate us from them, rise the mountains of Moab, purpled by the sunset, solemn with the glooms of falling evening. A double peak is pointed out as Mount Nebo.

There is light enough for us to read in the clear type of the "International Teachers' Edition," how "Moses went up from the plains of Moab unto the mountain of Nebo, to the top of Pisgah"—(or the hill), "that is over against Jericho. And the Lord showed him all the land of Gilead, unto Dan, and all Naphtali, and the land of Ephraim, and Manasseh, and all the land of Judah, unto the utmost sea."

("That is the Mediterranean," is interjected here.)

"And the south, and the plain of the valley of Jericho, the city of palm-trees, unto Zoar."

Jericho was worth having in that age, and fourteen hundred years thereafter, when enamored Antony deeded it to Cleopatra, from whom Herod rented it. The flow of milk and honey was scarcely a figure of speech. The fertile plains, irrigated from the Jordan, were grazed by flocks and herds, and bee-culture was one of the chief industries of the population. "A divine region," says Josephus, "covered with beautiful gardens and groves of palms of different kinds."

There are not a dozen palms in sight, when we stroll to the top of the bank beyond the camping-ground, for an unobstructed view, but the Jericho oranges have a reputation of their own. A man sent into the town on an errand brings back immense citrons, six inches in length and ten in circumference, sweet oranges, and, rather as a curiosity than to please the palate, sweet lemons, which we find insipid. It is David's will that the dining-tent be decorated this evening, and the messenger is laden with fruit-bearing branches of citron and lemon-trees, long sprays of acacia, studded with fluffy golden balls, deliciously fragrant; an unknown white flower, and, because I have expressed a desire to see them, bunches of the famous "Dead Sea apples." Several clusters are hung against the palankeen curtain, and Alcides transfers the group to a "film," while I cut open some of the nondescript vegetables for the purpose of classification. The writer who called them "a sort of potato," looked no further than leaves and flower, which strongly resemble those of the potato. But they grow upon a bush from two to

THE CAMP AT SUNSET.

six feet high, and in color and shape are identical with our yellow egg tomato. The seed, also, are those of the tomato, but dark-brown, as is the soft pulp surrounding them. Jamal says that they give forth a goodly smell at a certain stage of ripeness. Now, we find in them no particular odor of any kind beyond a certain rank weediness shared by them with scores of other plants. They are not edible, nor yet poisonous, and assuredly are never filled with ashes.

"What sort of growth do you suppose the poet had in mind when he wrote—

And Dead Sea fruit shall quench the thirst
Of those who long for wine?"

queries Alcides, returning the kodak to its case. "And no tourist in modern Egypt can find the genuine lotus. There is a sign in New York City: 'Destroyer of Moths.' Foreign travel might be advertised as a 'Universal Myth Destroyer.'"

"We are to have visitors!" David comes up to say.

A long man and a short woman are alighting from their horses. It does not surprise, however it may annoy us, to recognize the Inevitables. It is in accordance with the law of the general fitness of things—especially of human things—that Dr. and Mrs. Sharpe should have "dropped in" upon our sunset rest on their way to the Jordan Hotel in the lower valley.

DEAD SEA FRUIT.

With hospitable tact, the head of our commissary department has fresh tea made, and our easiest chairs are paraded for the visitors. It chimes in with our present opinion of the ex-college-president that he rejects sugar and cream, and gulps the boiling-hot liquid without winking. Setting the cup down with unnecessary stress, he summons David.

"Now, I dare say you have been telling these 'Innocents Abroad'"—in grimmest pleasantry—"that the contemptible 'run' of rain-water up there is the Brook Cherith, where Elijah was fed twice a day by the ravens?"

David's inclination of the head and slight smile are the soul of the Oriental's respectful forbearance.

"And being a latter-day dragoman, you have thought it incumbent upon you to hint at doubts of the ravens themselves? Some traveled fool—the most obnoxious species ot tool extant—probably a professor of theology—has rattled off a lot of stuff about the same word meaning 'ravens' and 'Arabians' or, as others have it—'merchants,' or still others—'Orbites'—the inhabitants of the neighboring city of Orb. Each and all of these efforts to explain away the miracle plainly recorded in the inspired Word is an insult to the Book and to the learning of those who translated it into the vulgar tongue."

Alcides has got hold of an idea.

"If the word translated 'ravens' can be *made* to mean any one of the three other things you have mentioned, why not give Elijah the benefit of the doubt, and relieve his memory from the obloquy of eating tainted flesh? Anybody who knows the raven's habits and tastes must prefer this reading. What more likely than that the Bedouins would have protected and fed the fugitive who threw himself upon their hospitality? They would do it to-day, wouldn't they, David?"

"Without a doubt, Captain."

"The question, young gentleman, is not what you, or I, or anybody else would prefer—"

"Now, doctor, dear, *why* will you dispute with everybody?" Mrs. Sharpe slips in, plaintively. "It is *so* much sweeter and more Christian to believe what the Church and the dragomans, who were born here and have *lived* here all their lives, tell us, than to be forever in the seat of the scornful, as one might say. As for Cherith—doesn't the Bible say *expressly* that it dried up, after a while, and what more likely than when the summer weather came on—and, as you have remarked," turning to Alcides—"the idea of tainted meat *isn't* pleasant to a refined person."

Her spouse jumps up and strides away to his horse, calling harshly to the dragoman that the sun is down and they ought to be off. Mrs. Sharpe finishes the tea she has had qualified with "two lumps, a *good* deal of cream, and a *dash* of hot water—*if* you please!" before rising to follow.

"The blessed man is a saint at heart, but a *little* impatient of contradiction," she murmurs oilily with her "good night."

We renew our laugh over the interview after dinner, when a company of friends from the hotel below walk over the intervening half-mile to pass the evening with us. There are two clergymen among the guests and both have fled for their lives, having dared to lean toward what we call "the Bedouin theory."

We are thankful nothing was said to the critic of Nebo or Pisgah, in looking by the light of a late moon toward the solemn height across the river. Already we have learned to seek out that point of the landscape, and dream, in gazing upon it, of the solitary man who climbed it to see "sweet fields beyond the swelling flood," before dying there alone.

CHAPTER XXVIII.

THE DEAD SEA.

THE march over the plains is begun betimes on the morrow It leads, shortly after quitting the camp, right through a Moslem burial-ground. Six men are hollowing a shallow grave in the sand; about a stone's cast away, the corpse that is to fill it, wrapped in coarse matting, is laid in the shadow of a thorn-tree; a group of women wail and sob intermittently about it; the intermission is devoted to cheerful chat, probably to gossip.

"It was an old man, or they would not be laughing and talking between their crying-fits," says David, in riding by. "If he were young, they would mourn loud and long."

Chalk hills bound the tedious level in the direction from which we have come; the purple mountains of Moab are upon the left; all about us are spread sand-dunes and clumps of scrub-growth; thorn bushes—the "Spina Christi"—or "Christ-thorn," with wild willows, and now and then a bunch of pinkish flowers growing upon a tall succulent stem, with fleshy leaves, and a shrub resembling the larch, but with more feathery foliage, that bears brownish-red berries. In an hour the land dips to a lower plain where the recent rains have settled in bluish mud. The horses keep upon drier ground, David and the sheik dashing away for horse-back practice upon a stretch of short, scant grass. Both are well mounted, and the racing, the sudden turnings and wheelings, the bounds and curvetings of the blooded animals are interesting to watch. My muleteers are ankle-deep in the sticky mire, drawing their feet out with a sucking sound. The subaltern has taken off his red slippers (it is always a puzzle to me how he ever keeps them on, for they are turned in under the heel and held in place by the toes only) and carries them in his hand till I suggest that he put them into the boot of the palankeen.

The sun shines hotly overhead by the time we get an unobstructed view of the Dead Sea. A blue haze hangs above it and veils the encompassing range of mountains; ineffable lights play over hills and sea; the water gives back the sunshine from dimpling waves. All the storied desolateness of the scene is to be found upon the land. The horses kick up salt as they walk, and about the hoof-prints in the sands are borders like hoar-frost—salt and so acrid to the tongue, that I marvel at the mules' disposition to snatch mouthfuls of it.

Hereabouts, archæologists incline to locate Sodom and Gomorrah, and not in the bed of the salt sea.

Another shock to early belief and dreams! Never more can I repeat the weird lines conned with delightful shivers in my girlhood:

> The wind blows chill across the gloomy waves
> How unlike the green and dancing main!
> The surge is foul as if it rolled o'er graves;—
> Stranger! here lie the cities of the plain.

Another myth goes down before scientific investigation. The waves are not gloomy, much less foul, and the tradition that the cities of the plain lie under-

WOMEN CROSSING PLAIN ON THEIR WAY TO THE BURIAL.

neath is—traditional! Loath to admit all this, we have fled to the history and the testimony concerning the destruction of the guilty towns, and are confounded that the main narrative and confirmative passages to which the margin refers us, do not so much as intimate the engulfment by the Dead Sea of Sodom, Gomorrah, Admah and Zeboim. Again and again, we read, that they were "overthrown by the Lord in His wrath," but the incoming of the bitter waters has no part in the tragedy. Furthermore—the "cities of the plain" would show that they were at

THE DEAD SEA.

the basin of the Jordan lying at the upper, or northern end of the sea. The southern border is sharply precipitous, and a line of mountains would have hidden from Abraham in Hebron, "the smoke of the country that went up like the smoke of a furnace," had this "country" occupied the present bed of the Dead Sea. Excavations in the mounds near the Jordan are constantly unearthing the rubbish of buried towns, and the subsoil of the waste lands over which we are riding is saturated with bituminous ooze. "The vale of Siddim was full of slime-pits"—*i. e.*, say erudite authorities, liquid bitumen, such as was used for mortar in the

SAND-DUNES AND CLUMPS OF SCRUB GROWTH.

tower of Babel, and by Semiramis in building Babylon. The doomed cities came to their ruin by brimstone and fire from the Lord. The whole region would burn like tinder if once ignited. The agency of water in the terrible work does not appear in the sacred record.

"I am afraid we must give it up," concludes my fellow-student, ruefully. "But should we encounter Dr. Sharpe here—(which may the fates forfend!) we will not admit our discomforture to him. He is quite capable of discounting the whole story, Bible or no Bible."

Two large boats are beached upon the shelving shingle, lapped by the sparkling waves. The government has a monopoly of the bitumen that floats upon the surface of the Dead Sea, and the boats are here to collect it. A piece weighing fourteen tons arose to the top of the water last week, and smaller fragments are found in abundance, especially after a storm of wind. To the government alone belongs all the salt brought into and sold in Syria. The drift-wood along the shore of this sea is thickly encrusted with it, and the peasants of the region come by stealth to scrape it off for family use. When detected in the act, they are arrested and punished as thieves.

Of course, Alcides must take the regulation bath in the saline waves, so buoyant that, as he reports afterward, it is impossible to sink oneself without great effort, and then the body bounds back to the surface like a rubber ball. While he is absent with an attendant, I pace the wet shingle, as hard as a floor, and find mood and scene wondrously tranquil—not at all what might be expected with the Dead Sea washing the tips of my boots, and the overthrow of Sodom and Gomorrah vivid in my mind. Far away toward the southern extremity of the shining sheet, which is likewise known as "Lake Asphaltum," once stood Machias, in whose dungeons John the Baptist was murdered, and nearer to us "little Zoar," spared that Lot might take refuge there. The latter place is yet more to me because it was the boundary of the sweep of the undimmed eye of the prophet, nigh unto death, and yet in the full vigor of manly perfection.

My eyes return, and lingeringly, to "Nebo's lonely mountain," sublime in noon-tide serenity, still the warder of the plain and sea. All the beauty of the "city of palm-trees" and the "glory" (not the "swellings") of Jordan, could not have hindered a compassionate thought from crossing the seer's mind as he beheld from afar the region he had so lately described as "the land of brimstone, and salt, and burning, that is not sown, nor beareth, nor any grass groweth therein—Sodom and Gomorrah and Admah and Zeboim, which the Lord overthrew in His anger, and His wrath."

ENCAMPMENT UPON THE JORDAN.

CHAPTER XXIX.

THE JORDAN.

THE luncheon-tent is pitched upon the very spot from which, says tradition, Elijah ascended to heaven.

Stretched luxuriously upon the rugs with which David has carpeted the turf, we, as is our custom, read the story of the event from the original record, and draw conclusions unbiased by popular tales and erudite commentators.

The two prophets passed over Jordan between the waters divided hither and thither by the stroke of Elijah's mantle;—"And it came to pass as they still went on and talked, that, behold, there appeared a chariot of fire and horses of fire, and parted them both asunder; and Elijah went up by a whirlwind into heaven."

"Elisha took up also the mantle of Elijah that fell from him, and went back and stood by the bank (or 'lip,' says the margin) of Jordan."

We are within fifty yards of the shelving brink, a distance scarcely worth the mention of a return to the river. It is more likely to our imaginations that the stupendou miracle—second only to the Ascension of the King of Glory and the lifting of the celestial gates to let Him in—took place much further away toward the interior, perhaps upon the bushy plain back there, over which horses and mules have forced a way through thicket and briery tangle to reach the sacred stream.

All visitors agree in pronouncing the Jordan a disappointment at first sight. It is, to put it plainly, a mere creek—by actual measurement, one hundred and ten feet wide, and seemingly not more than fifty. So delusive is the apparent width that we cannot credit the dragoman's respectful insistence upon his figures until the picture taken by the kodak of the thither bank shows, by the blurred lines, that the jungle of bulrushes and willows lining the water's edge is almost beyond range. Nor is there anything dignified in the environing scene. A fringe of ragged trees, tall flags and flowering reeds is broken in front of our noon encampment by an open space through which we view the Moabitish shore. On that side the ground rises somewhat abruptly and is more heavily wooded. Back of it are the mountains of Moab—mysterious barriers which we long hourly the

A FERRY OVER THE RIVER JORDAN, SHOWING THE THITHER BANK OF THE JORDAN.

more intensely to visit. In our immediate foreground, unconscious David, water-jug in hand, has stopped to offer friendly admonition to the younger dragoman of a party of tourists who occupy a conspicuous position in the field of the camera, and are lunching among the willows. In common with many of his guild, the young fellow is exalted above measure by the circumstance that he can speak both English and Arabic, and that the foreigners he has in charge are at his mercy in the matter of topography and bargains. He is sharply authoritative in forbidding a girl of fourteen and a boy of twelve not to cut canes in the bushes up the stream.

PARTY OF TOURISTS ON THE JORDAN.

"Don't go ten steps away unless I give you permission!" is the order that catches Jamal's ear.

"I have told him that he had no right to use that tone in speaking to any of his travelers," David answers our interrogations as to the meaning of the scene. "He is young and 'fresh.' In time he will learn better. No! there is no real danger in wandering off a little way from a party. That is, not now. In the old time, robbers were hidden in the brushwood and canes."

Alcides, accompanied by Serkeese, disappears among the undergrowth lower down the bank, intent upon a literal seven dips in the Jordan. A lesser number would not rid skin and hair from the smart and stickiness left by the Dead Sea bath. The men of the party under the broken willow tree, taking the children along, stray off to cut walking-sticks as souvenirs of day and spot. I send one of the rugs furnished by the more experienced dragoman to the one lady of the company under the escort of the "fresh" young Syrian, when I see her lie down upon the sloping turf, and double her arms under her head for a pillow. Then, among my cushions, I look up "Jordan" in the "Biblical Gazeteer" at the back of "the best Baedeker for Palestine."

While reading of the passage of the twelve tribal representatives through the "dry ground" of the river-bed, each with a stone upon his shoulder, "which they took out of Jordan and carried them over with them unto the place where they lodged, and laid them down there," finally at Joshua's command, "pitching them in Gilgal,"—the rush of the muddy, crooked stream in the stillness of high noon takes on solemnity to the ear of my spirit. I close the book upon my finger and look from the shadow of the tent-hood upon the historic current as one awakened out of sleep. Hereabouts, beyond the shadow of doubt, the host of Israel crossed into the Promised Land. Midway in the stream—perhaps upon the very spot on which I am gazing—the ark, upborne by the white-robed priests, held back the flow from the far-off source at the foot of Hermon "until all the people were passed clean over Jordan."

"UNCONSCIOUS DAVID, WATER-JUG IN HAND."

Slowly, and as if I had never heard or seen it before—I reread the wonderful tale:

"And as they that bear the ark were come unto Jordan, and the feet of the priests that bear the ark were dipped in the brim of the water (for Jordan overfloweth all his banks all the time of harvest), the waters which came down from above stood and rose up upon an heap very far from the city Adam, that is beside Zaretan; and those that came down toward the sea of the plain, even the salt sea, failed and were cut off; and the people passed over right against Jericho."

Muddy, tortuous, and neither broad nor long, the Jordan has yet a certain dignity of its own, or so it seems to me, as I watch the swift, steady flow toward the salt sea that is to swallow it up and render no account of the flood poured into it. At this point it is nine or ten feet deep, and, as Alcides reports upon his return, icy cold. He had meant to land upon the Moab side after swimming across, but hastened back, fearing cramp should he remain too long in the water. We are thankful for the change of purpose upon hearing that the rushes of the opposite border mask a most deceitful and dangerous quagmire. The river must have undergone many and mighty changes since the "this day" (B. C. 1427) in the which, we are told, other twelve stones, "set up by Joshua in the place where the feet of the priests which bare the ark of the covenant stood," were known to be there. They are probably there now, buried deep beneath the alluvial deposit of centuries.

OUR SHIEK AT THE JORDAN.

The longer we look at the river and talk of the "wonders" done in and about it, the more majestic it grows. There is silent eloquence in its swift progress to the sea of death, grave, self-contained consciousness of the part it has played in the world's history, and of its importance as a type of the Dark Stream each must ford for himself when the word is given to "pass over Jordan."

Over yonder were pitched the tattered tents of the great host, tremblingly expectant of the entrance denied to their sinning fathers into the Canaan of prophecy and prayer. We may have in sight the place from which Joshua issued

"GATHERED BY DAVID FROM THE EDGE OF THE STREAM."

the command;—"Sanctify yourselves; for to-morrow the Lord will do wonders among you."

The morrow that was never to dawn for the sleeper,—

> "On Nebo's lonely mountain,
> On gray Beth-poor's height."

Again, and from the deeps of our hearts—"Poor Moses!"

Not far from where we are sitting, John the Baptist preached, and "there went out to him Jerusalem, and all Judea, and all the region round about Jordan and were baptized of him in Jordan, confessing their sins."

Above our very heads the "heavens were opened, and" the Divine Man "saw the Spirit of God descending like a dove, and lighting upon Him." The exact spot cannot be designated with any degree of certainty, although since the fifth century after Christ, and we know not how much earlier, crowds of pilgrims resort yearly to the region round about Jordan to be baptized and to bathe in waters they esteem as holy.

We have left no other place in Palestine with more reluctance than we acknowledge as, having seen our bottles of Jordan and of Dead Sea water bestowed in the bottom of the palankeen, and a bundle of walking-sticks and feathery grasses gathered by David from the edge of the stream, strapped upon Serkeese's luggage, we turn our backs—probably forever—upon the scene that looks so tame and is fraught with so much of supreme and tender interest. Our little caravan is headed straight for a clump of trees marking the sight of Gilgal—Joshua's Gilgal. The oasis of the Jordan dwindles into briers and thorns and such coarse herbage as can thrive in the salty soil, before we reach the few mounds—or rather hillocks—supposed to mark the site of the place where the rite of circumcision was renewed in the temporary lull of hostilities caused by the awe cast upon the people of the land by the news of the miraculous crossing of the Jordan. Where, too, the children of Israel were encamped when they kept the passover on the fourteenth day of the month at even in the plains of Jericho.

HOUSE OF ZACCHEUS.

"And the manna ceased on the morrow after they had eaten of the old corn of the land; neither had the children of Israel manna any more, but they did eat of the fruit of the land of Canaan that year."

PLAIN OF JERICHO.

There were two millions of the chosen people by now. The land to which they had emigrated had need to be all aflow with milk and honey, and stored beyond our power of conception as we traverse the waste places, with "butter of kine, and milk of sheep, with fat of lambs, and rams of the breed of Bashan, and goats, with the fat of kidneys of wheat, and the pure blood of the grape."

We skirt modern Jericho upon our campward way. It is a dismal collection of mud-huts, covered with turf and with reeds packed over with earth. A ruin of brick walls, unroofed and crumbling away, is pointed out as the house of Zaccheus. Beyond is a modern hotel, insignificant in size, and mean in architecture. In following the winding, muddy lanes, we have glimpses over a wall of a garden belonging to a monastic order, wherein grow lemon, orange and citron trees; the feathery foliage of the acacia-tree brushes our cheeks in the narrowest turnings, and we secure handfuls of the little golden tufts to be used as perfume in handkerchiefs and glove-cases.

It has been a long and an eventful day, and like the flash of lights from the windows of home, falls upon our vision the gleam of color from the peak of white tents, when the muddy maze has been threaded. A breeze from the hills leading up to Jerusalem almost four thousand feet higher than the plain of the salt sea and the plain of Jericho, lifts into distinctness the crescent-and-star of the Turkish flag over the dining-tent, and the stars and stripes under whose dear folds we are to sleep to-night.

CHAPTER XXX.

SUNDAY IN CAMP.

THE phrase "Sabbath stillness" borrows new meaning from the experience of a Lord's day passed in a wilderness camp. The very mules comprehend that nothing is expected of them, and do not offer to rise from breakfast until luncheon-time. The big leader of the palankeen team lies in the shadow of the burden he is accustomed to bear six days in the week, in lordly forgetfulness of yesterday and carelessness of to-morrow—a sedate preacher to the care-taking bipeds who watch him from the tent-door.

Our morning-reading has to do with Joshua and with Jericho. The city of palm-trees, and of roses rivaling those of Damascus in prodigality of bloom, was still a place of note in our Lord's day, and the great mound to the left of our camp probably hides the ruins of one of the fortified towers from which, perhaps, as we like to think, the inhabitants, "straitly shut up because of the children of Israel," watched during six suspenseful days the daily circuit made by priests in sacerdotal robes, bearing the ark, and followed by "all the men of war." Fancy dwells, in the perusal, upon the awful silence that prevailed among a host so vast that it must have girdled the place, unless measurements made in modern times err as to the extent of ancient Jericho.

"Ye shall not shout, not make any noise with your voice," thus ran the leader's order—"neither shall any word proceed out of your mouth, until the day I bid you shout. *Then*, shall ye shout."

"And it came to pass on the seventh day that they arose about the dawning of the day," I read aloud.

There was need of an early start and brisk work, for between that gray day-break and the going down of the sun behind the hills lifting their bald heads against the cloudless heavens to-day, the town was compassed seven times. It would be well on towards evening when "it came to pass when the people heard the sound of the trumpet and shouted with a great shout that the wall fell down flat"—or, as the margin has it, "under it." That is, at the mighty roar which must have reverberated from distant Nebo and shaken the waters of the salt sea.

In the sanguinary history of conquest which succeeds the tale of this marvelous siege, we come once and again upon the war-cry—"As Joshua had done to Jericho and her king."

Alcides calls a halt, by-and-by. In his opinion the Jews were never a war-like people unless urged to the front, and then required divine interposition in every exigency.

"Who of us does not?" I interpolate here.

"True—but listen! The surprise of Ai was planned by the Lord. When Joshua made a forced night-march from Gilgal, we are expressly told that "the Lord discomfited" the five kings from Jerusalem, Hebron, Jarmuth, Lachish and Eglon, so that they fled before Israel, and "He cast down great hailstones from

"IN LORDLY FORGETFULNESS."

the heaven" upon them in the retreat, so that "they were more that died with hailstones than they whom the children of Israel slew with the sword."

The young critic breaks off to observe,—"It makes one dizzy to think of it. And how kings quarreled for the possession of this miserable stretch of waste land. Eglon of Moab 'possessed the city of palm-trees,' and David's shorn ambassadors took refuge here until their beards were grown, and three hundred and forty-five former inhabitants of Jericho were returned to the beloved oasis after the captivity; Antony thought it good enough to give to Cleopatra; Herod the Great coveted and bought it from her, fortified and built in it his favorite palace—where he died by the way. There was a big hippodrome here, in which he ordered all the leading citizens of Jerusalem and the vicinity to be imprisoned when he knew his end was near, and left a dying injunction that they should be massacred as soon as the breath was out of his body—a Shylockian device to procure a general mourning for the dead king. And look at Jericho now!"

The forenoon wears on peacefully. The writing-table is brought out to the tent-door and—always in sight of Nebo—letters are written to the dear ones in the distant home—all done as in a dream-world, so steeped are thought and fancy with the tragic story of that dead past.

We are still dreaming as we visit what modern travelers and natives know as the Sultan's Spring; a living fountain bubbling from the side of the mound of

MODERN JERICHO.

which I have spoken as covering a fortification of the Jericho destroyed by Joshua. Excavations have brought to view broken marbles and pottery, the remains of temples and dwellings belonging to a long-buried era. The whole hill is a conglomerate of dumping grounds of different ages, kings and conquerors having acted as scavengers. A bit of reddish stone picked up at random on the hillside proves upon inspection to be a fragment of *rosso antico* marble, susceptible of exquisite polish, and is laid away to do duty as a paper-weight, when Jericho and the unreal outlying lands shall have taken their place among the other by-gones of this witching winter.

The Sultan's Spring has another name in David's mouth.

"If the Captain will look at Second Kings, second chapter, from the nineteenth to the twenty-second verses," he intimates as we stand above the shallow pool, and remark upon the solidity of the masonry securing and bounding it.

It is the well-known incident of the appeal of the men of Jericho to the prophet tarrying in their midst, after the translation of Elijah while the sons of the prophets sought vainly for his vanished friend. For, "Peradventure the Spirit of the Lord hath taken him up, and cast him upon some mountain, or into some valley."

IN THE TENT DOOR.

"The situation of the city is pleasant," the Jericho people represented, "but the water is naught and the ground barren." * * * "And he went forth unto the spring of waters, and cast the salt in there and said, Thus saith the Lord, I have healed these waters. There shall not be from thence any more dearth or barren land. So the waters were healed unto this day."

"This is Elisha's Fountain, if wise men are to be believed," comments the guide.

The healed waters are crystal-clear, and, led off in various directions from the outlet, make green and flourishing the small gardens of herbs in the neighborhood. The sides are covered with succulent plants. All growing things within reach of it press eagerly toward the life-giving stream.

"It is a pity that Elisha set the she-bears upon the children so soon after a deed the good of which has come down to us," regrets the reader, and I cannot gainsay the stricture.

In flat contradiction of a serene sunsetting, the eastern sky is a passion of flame color as I draw aside the flap of my tent-door on Monday morning. Against it lie the mountains of Moab, misty purple as an August plum; waves of softest pink quiver up to the zenith; the yellow sand-hills catch the reflection and deepen into orange-red. In all the camp no one is astir except John and the head-muleteer; horses and mules have not aroused themselves from their Sabbath rest. The world enclosed by the eastern and western ranges of hills is still and cool, and transfigured by the magic bath of glorious color. Hastily throwing on a dressing-gown, I steal to Alcides' door and summon him to enjoy the sight with me.

ELISHA'S FOUNTAIN.

Presently we repeat in unison, "It will be foul weather to-day, for the sky is red and lowering."

David is of like opinion for once with Pharisee and Sadducee, for he hastens the preparations for departure. The cords and stakes of the dining-tent are loosened while we are at breakfast. When we issue from the shaking tenement, palankeen and horses are ready. The sunlight has not touched the lower valleys when we take the road to Jerusalem. The defiles gloom blackly between the bleak

FOUNTAIN OF ELISHA, SEEN FROM MOUND COVERING ANCIENT JERICHO.

heights; the profound stillness is even more oppressive than when we traversed the same road last week. Above the sullen mists that hide the depths play broken rainbows; the bold brows of the distant hills are wreathed with fogs, turbans of white and gray, changing to the prismatic colors as the sun darts through a rift in the gathering clouds. The mules pick their way among rolling stones, and along shelving paths, where the palankeen tilts crazily; David and the sheik ride mutely ahead, heads lifted often toward the darkened sky.

Alcides alights to unstrap a roll from Serkeese's luggage. It contains mackintosh, rubber boots and helmet, and is swung now from Massoud's saddle. At

PREPARATIONS FOR DEPARTURE.

the dragoman's suggestion, I get out my own waterproof, and fasten the palankeen windows. Our preparations are made none too soon. As we turn the pass beyond the Khan of the Good Samaritan, the storm strikes us, sharp and sudden. We have hardly time to see it rushing toward us in slant, white sheets, from a dozen different directions, when the road is filled with a falling volume of water, like a cloud-cataract, under which the horses stagger and the mules stop for a blind moment. There is a steady, bitter rain in our faces all the rest of the way. Enveloped in waterproof, furs and rugs, I sit well back in my corner, and, but for the spray drenching cap and veil, am unharmed. Serkeese pulls his abieh over his head, and bends almost double; Alcides and David ride erect, the rain pouring in sheets from their glistening mackintoshes, the eyes all that is visible of their faces under the visored helmets; the sheik is as impassive as a mummy. The foremost muleteer, at the first dash of the flood, has snatched off his heelless red slippers, and tucked them into the foot of the palankeen. In three minutes there is not a dry thread on him; in five, every wet thread is a tiny aqueduct, every fold of his garments a pool. The rain is at its bitterest worst when he breaks into song—the nasal drone that passes with the Syrian peasant for music—and keeps it up at scant intervals for at least five miles.

SERKEESE AND LUGGAGE.

I seem to have listened forever to the dissonant chant by the time we halt for luncheon at the ruined khan opposite the Apostle's Fountain. At one end an arch of Roman masonry protects the pallet and rugs laid upon a ledge on which the proprietor sleeps at night. The roof is of reeds overlaid with turf, and the rain drips through crevices upon the earthen floor. Six or eight men huddle about a fire of thorns built in the middle of the room. They fall back at our approach, and after David has laid our luncheon upon a rude table at the back of the room, nobody glances in our direction. It is a taciturn group. The cold and wet have taken all the spirit out of them, and there was never much. They part their ranks again to let a black woman pass to the fire, with a baby in her arms. The blackest, forlornest baby I have ever beheld, the creature sits upon a stool drawn up into the very area of flame and smoke, and shivers piteously in a violent ague, never uttering a whimper, but staring solemnly at the fire.

"It will die, I think," says the mother, stolidly, to David, who replies, "Indeed, I think so," with more show of concern than is manifested by her.

At the parting of the ways between Bethphage and Bethany, the sheik rides close to my palankeen, thrusts in a brown hand, and utters the only English word I am to hear from his lips:

"Good-bye!"

"His village is near," explains David.

I have been in a deep reverie for a long hour. Thoughts of the Homeless Man who walked this weary road, foot-sore and heart-sore, have well-nigh supplied

OUR HEAD MULETEER.

the place of sight. As we pass Bethany, dimly visible through the curtain of rain, the solitary Figure is yet more vividly present to my imagination.

In the exaltation of spirit begotten by these musings, it is a shock when the palankeen swerves abruptly to the side of the road to get out of the way of a carriage and three horses driven abreast, in Syrian fashion, and coming down the hill at full speed. Before I can ask a question, the vehicle stops in mid-career, the horses falling back upon their haunches, and out springs Domian, the ex-partner of David Jamal, his dark, handsome face full of kindly solicitude, an umbrella in one hand, a bottle of brandy, already uncorked, in the other. In a trice we are bestowed among dry rugs in the interior of the coach, hot-water bottles at our feet, and the three horses are galloping back to town.

"Mr. Gelat and I thought you should be met, madame," says the dragoman, and we discover, all at once, that we are drenched and chilled to the bone, and must have been utterly wretched in another half-hour.

"I shall have a fire kindled in my room!" I articulate between chattering teeth in alighting at the door of the Grand Hotel.

Domian says nothing, but hurries me along corridors and across a great hall to my familiar quarters, throws the door open, and with the current of genial warmth there streams out upon us the red glow of firelight and the fragrance of tired woman's sweet restorer,—a cup of hot tea.

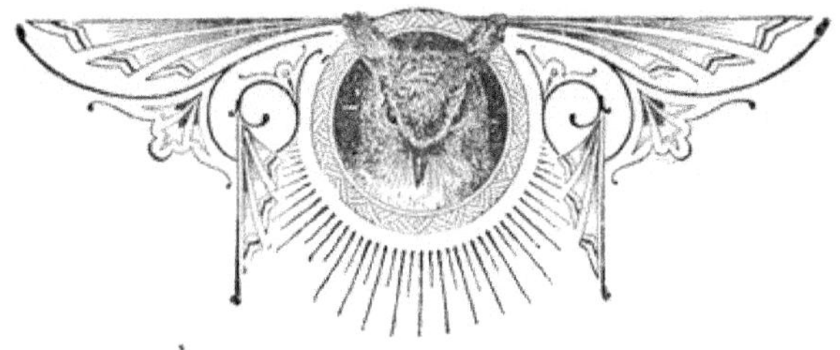

CHAPTER XXXI.

"THINE ANCIENT PEOPLE THE JEWS."

UNDER this name I heard them prayed for every day throughout my infancy and girlhood. The phraseology in which the patriarch of the household remembered them at the family altar varied little in all those years:

"We pray Thee to have mercy upon Thine ancient people the Jews, and bring them into Thine Everlasting Kingdom, together with the fullness of the Gentiles. May they look upon Him whom they have pierced, and acknowledge Him as King of kings and Lord of lords."

I am saying it over to myself on this murky afternoon, as, despite the corrugated soles of my overshoes, I slip and slide upon the greasy mud of the Jewish quarter in the City of David. The smells are the foulest I have encountered, and we stir them into aggressiveness in our passage through the damp, breezeless air. It is Friday, and marketing is lively, in preparation for the national Sabbath. Every huckster has put forward his choicest wares; women, conspicuous by their uncovered faces, have baskets on their arms, and haggle shrewdly over "green stuff," groceries and meat. Once we are shoved against a wall by a crowd of both sexes gathered about an auctioneer who is selling tainted fish at a piastre (eight cents) a pound. The ground is wet and steep; the stones are treacherous, the throng motley and unsavory, with an indescribable air of sordidness evident through the squalor of the region.

My companion is Rev. Joseph Jamal, a cousin of our educated dragoman, and assistant rector of the English Church in Jerusalem. Under his guidance I have seen the various benevolent institutions established in Jerusalem by the Church Missionary Society which is an honor to English Christians and philanthropists. The Ophthalmic Infirmary, where tens of thousands of sufferers from the fearfully prevalent diseases of the eye are treated annually; the Dispensary, well stocked and admirably managed; the Industrial School, where carpentry, printing, bookbinding, etc., are taught; the Hospital, fifty-five years old, which will be, ere long, transferred to the fine new building now in construction outside of the walls—one and all are fraught with deep interest to her who has seen enough of the Jewish population of Palestine to appreciate the needs and the discouragements of the work undertaken by the splendid organization named above. Mr. Jamal's twenty years of labor in the service in the Church Missionary Society have fully qualified him as an intelligent and trustworthy authority upon the subject which, during the past few days, has especially engrossed my time and thoughts.

Threading the maze of filthy streets, we arrive presently at an archway so low that my tall guide stoops low, and I have to bow my head in passing it and entering a sort of tunnel, wetter and fouler than the open street we have left, and sloping downward. Another turn, and a plunge of several steps, and Mr. Jamal knocks at a low, dingy door in a blank wall. It is opened by means of a cord running along the dark passage, and we see nobody until we are met at the head of a flight of unclean stone steps by a kervasse in dirty uniform, with a red fez upon his head. He takes us into an apartment of fair size, the upper half raised by two steps above the lower, and lined on three sides by a cushioned divan, on which we are seated. This is the drawing-room of the Chief Rabbi of Jerusalem. The floor is covered with matting; a few rugs are scattered here and there, and two small stands are set back against the wall. Besides these there is no furniture. The kervasse, having taken our cards to his master, returns in four or five minutes, ushering in two old men. One, tall and

IN THE "BOX COLONY," JERUSALEM.

with traces of former dignity and comeliness, walks in advance of his companion, who is his inferior in appearance and in office. Both wear dark-blue cloth gowns, lined and edged with coarse fur, and bands of like material trim the caps, which are round and flat on top. A moment later, a third man, similarly attired, enters, unannounced, and sits down upon the divan with the others, we facing them from the other side of the room.

After a few and ceremonious preliminary remarks, we come to the chief object of my call. I ask the master of the house through my interpreter if he attaches

"A MAZE OF FILTHY STREETS."

any significance to the influx in late years of the Jews from other lands into Palestine. Also, if he can give me an approximate idea of the number who have thus immigrated within ten years.

"If you would know how many have come in the past sixty years, I should answer that there were but one thousand Jews in Jerusalem and the vicinity in 1833. There are thirty thousand now."

"How do you account for the steady increase of immigration? Are all drawn by the same motive?'

"The Jews come to Palestine because they love it as the land of their fathers and their own country. Some have come expecting the Messiah. They are foolish. When He comes, He will rule the whole earth, not merely this little corner of the globe."

"You expect, then, His personal advent? What will preface it?"

JEWISH IMMIGRANTS IN JERUSALEM.

"The Great Fight of Armageddon must come first. Gog and Magog will appear and be overthrown. There will be a terrible bloody conflict of all nations in the Valley of Decision."

"Where there is not room to deploy four regiments!" I heard Dr. Merrill say indignantly, last night, but I forbear to quote the sensible objection to the chosen battlefield. I ask, instead:

"Where is the promise of His coming? Do you see signs of the approaching gathering of nations?"

"Who can say? The political horizon is dark, and may mean much. Since the prophets passed away there is no man who can read the signs of the times."

"Where will the Messiah first appear?"

"He will descend from heaven upon Mount Safed, the highest point of Galilee. So say the holy writings."

(I recall that Safed is pointed out as the "city set upon a hill" to which our

Lord, ever ready to illustrate His teachings by natural and present objects, may have pointed in the Sermon on the Mount.)

"But," I say aloud, "we are told by Zechariah that when 'the Lord shall go forth and fight against those nations as when he fought in the day of battle, His feet shall stand upon the Mount of Olives which is before Jerusalem on the east.'"

"True. The Messiah will proceed from Safed to Olivet."

"I read further that the Mount of Olives shall 'cleave in the midst thereof toward the east and toward the west, and there shall be a very great valley, and half of the mountain shall remove toward the north, and half of it toward the south.' Will this prophecy, in your opinion and in the opinion of other learned men, be literally fulfilled?"

"HIS INFERIOR IN APPEARANCE AND IN OFFICE."

My host inclines his head in grave assent. The colleague who sits next to him says decidedly, "Certainly, no one has ever questioned it."

"Where—may I ask—do you read the prophecy concerning Safed?"

"In the Talmud," with the air of a disputant ending a controversy.

But I am intensely interested, and my interpreter being altogether in touch with my mood and desire, conveys my meaning so faithfully that I cannot refrain from further researches.

"Tell him," I say to Mr. Jamal, "that some of the most learned Rabbis in America no longer expect a personal Messiah. They believe that the prophecies relative to His coming point to the perfectibility of human nature; to an advanced state of morality and subjugation of whatever is base and vicious in man's nature and conduct; to the cessation of war and crime, the elimination from body and mind of all that engenders sorrow, pain and death itself."

For the first time, the old man gives signs of excitement as this speech is translated to him. He crosses one leg over the other nervously; his black eyes gleam under the white brows; his raised hand and voice shake with agitation.

"No devout Jew believes such a monstrous thing! The men who assert it are infidels—materialists. The Messiah will be a real personage, great, holy, powerful, perfect, and He shall reign in the Mount Zion, forever and ever."

I return to a former question:

"When will he come? Are there indications that the time may be near?"

My venerable interlocutor retires unequivocally into his shell of dignified and official reserve.

"Who can say? That is in GOD'S hands—not in mine."

It is obvious that further catechizing would be unwelcome, and having partaken of the usual refection of sweetmeats and coffee, we exchange a few conventional compliments and part amicably.

Our next visit is to a poorer Rabbi, living in a more lowly abode, but as genial as the former was politely-frosty. He belongs to a sect whose business is the study of the law, and the shabby room is surrounded with bookshelves. So far from eluding such queries as I have put to his superior in office and worldly gear, he talks enthusiastically of his belief that the Kingdom of the Messiah is near at hand. He holds the same view with the Chief Rabbi as to the Great Battle of Armageddon.

"Gog and Magog are, I am inclined to think, Russia. All nations will be engaged in the Valley of Jehoshaphat. The right will conquer, the God of Israel fighting for it. A congress of nations will be held and decide to restore Palestine to the Jews, who will thenceforward possess it and cause the waste places to break forth into singing, the desert to bloom as the rose."

"But there is not room in Palestine—or in all Syria, for that matter—for one-half of the Jews now alive upon the earth."

He smiles benignantly and with the calmness of his convictions:

"You forget that they have never yet had all the Promised Land—'from the river of Egypt'—the Nile—'unto the great river, the river Euphrates.' The promise is 'ordered in all things and sure.' The whole world will then be at peace; nations shall learn war no more. All will worship one only and true GOD, the GOD of Israel."

I put out my hand impulsively and we shake hands cordially upon this.

"You are a Protestant!" I declare.

"We serve the same Lord," he answers.

After more sweets and more coffee, served by the Rabbi's wife--a motherly body—we are conducted by him into an amazingly small synagogue, one thousand years old, entirely underground, having been built when Jews were forced to worship in secret. It is lighted from above by means of two grated windows, like "man-holes," let into the pavement. There is light enough to enable us to examine a curious old manuscript copy of the law, over six hundred years old, brought from Bagdad and, at my request, the Rabbi reads the lesson of the day from it, a quavering intone, such as we have heard at the Wailing-place.

"It is a poor place," he says, running his eye around the dank den, "and must always have been very dark."

"Daniel prayed in a darker," I remind him.

His eyes twinkle, good-humoredly.

"And Jonah in still less desirable quarters!" is the unexpected rejoinder.

CHAPTER XXXII.

"THE BOX COLONY."

OUR excursion to-day has brought us to a muddy common outside the walls of Jerusalem. Right in the centre of it sprawls the most miserable village that can be imagined. Houses of unbaked clay and stubble, of cobble-stones held unsteadily in place by dried mud; board shanties roofed and sheathed with tin cans beaten out flat and nailed on, all one-roomed huts, built along miry alleys, hardly six feet wide—make up the "Box Colony" tenanted by immigrant Jews from all quarters of the globe. The ground is loaned to them rent free by a wealthy Hebrew resident of Jerusalem. Most of the rooms are windowless, and every door stands wide open to admit the light of a short winter afternoon. A bundle of rags, or a heap of straw, does duty in each as a family bed; braziers of charcoal are kindled with thorn-bushes, a heap of which lies in a central shed. There are children! children! everywhere. Four of us women have driven out from town as close to the settlement as a carriage can approach, then walk down the clayey slope. Not far from Dr. Sandreczky's hospital is a neat dwelling inhabited by two American missionaries—ladies, by birth and

"ALONG MIRY ALLEYS."

breeding, who, at their own charges, have devoted time and labor and life to the work of ameliorating the physical ills and enlightening the souls of this Jewish settlement. I have begged the favor of their escort, assured that the sight of these ministers of mercy will admit and recommend me everywhere. Mrs. Jamal accompanies me as interpreter.

Our first call in the forbidding circuit is upon a family of Aymonites, or Arabian Jews, usually esteemed as the most devout of all the sects. Our missionary friends have spoken warmly of their faith in God and the revelation made in their Scriptures of Him and His purposes toward their race. An elderly woman sits flat upon the mud floor, stitching at a nondescript garment of many-colored rags. Near her stands a striking figure—a man with an Arab face and head-gear. His eyes glow like live coals, his manner of greeting us has a gentle courtesy out of keeping with his patched abieh and bare feet. He holds a baby in his arms, who clutches his beard for protection while staring at us. We have not talked a minute before the room begins to fill with interested auditors. Every woman has a child in arms, and presses to the front; the few men skulk in the rear of the crowd, and peer in at the door; the children fill up the chinks in the living wall.

"CHILDREN! CHILDREN! EVERYWHERE."

The picture is peculiar and impressive. Near the door, Miss Dunn, of New

York, small in stature, with delicate features, dressed simply in black, takes in every feature of the scene through grave, pitying eyes; at my side Miss Robertson, a native Kentuckian, whose Southern intonations sound strangely and sweetly familiar to me in this far-distant land, salutes each new-comer with a smile or word, and when the conversation opens, hearkens with eye as with ear. Mrs. Jamal, handsome and vivacious, ready with both the languages in which the colloquy must be carried on, her izzar fallen back from her head, and the blue-flowered mendeel lifted from her face, is at my other hand. To the man, as a leader and a teacher among his people, my queries are addressed:

"SOMETHING LITTLE BETTER THAN BEGGARS."

"Where was your home before you came to Jerusalem?"

"In Arabia."

"What brought you so far from it?"

"We came as pilgrims, as Abraham of old, to the Land of Promise. Jerusalem is the City of the Great King. Our fathers builded it. It is our city."

"How have you fared here?"

"Badly enough, as you see. We left a land where we were comfortable, and had enough to eat and to wear, to become something little better than beggars."

"Was that wise? Do you not regret it?"

"Not for a moment. We bear all hardships patiently, expecting a release from captivity. Weeping may endure for a night. Joy cometh in the morning."

"You expect the Messiah to come before long—perhaps?"

A gesture of amazement.

"Who does not? The Deliverer will come to Zion. We are here to wait for Him."

"When will He come?"

He spreads out his hands in Oriental (and Hebraistic) fashion.

"Ah! who can tell? We Arabians have three proverbs—"Who can tell when the rain will fall? Who can foretell when a child will be born? Who knoweth when Messiah will appear?"

Mrs. Jamal's face lights up archly; she takes a step forward and answers quickly:

"But there are, in two of these cases, signs which we may read aright. When clouds gather, we say, The rain is at hand. When pain takes hold of a woman, she knows that her hour is at hand. Do you, who watch and expect, see no signs that the day of the Lord is at hand?"

"We believe that we do. I name but one. Houses are rising within and upon Jeremiah's measuring-line—'from the tower of Hananeel unto the gate of the corner, and upon the Hill Gareb and compassing about to Goath.' Have you not read that 'the whole valley of the dead bodies, and of the ashes and all the fields unto the brook of Kidron unto the corner of the horse-gate toward the East shall be holy unto the Lord'—that is, a part of the holy city? And beyond the Jaffa gate, behold houses upon houses, building, building continually. It is written that there shall be one great beautiful city stretching from Jerusalem even unto Jaffa."

His air is that of an inspired seer, and he piles word upon word breathlessly. A low chorus of what sounds like "Amen!" arises from the listening women. One kisses her baby convulsively and begins to sob. Tears are on other cheeks.

"Ask him"—I request of Mrs. Jamal—"where he has read the prophecy about the line of houses from Jerusalem to Jaffa?"

"In our sacred books," is the reply. "Not in the Scriptures that the lady knows."

"Will Messiah be born as child, or as a man?"

"He will come as a King, descending from heaven, and clothed with majesty and, as we believe, very soon."

"Will your children probably see Him?"

A sudden look at the unconscious infant, who still plays with the father's matted beard, a closer clasping of the little form, and he shows us a face from which the light of holy exaltation has faded into quiet resignation.

"Who can know that? God's ways and God's times are past finding out."

A woman breaks in excitedly here. Mrs. Jamal turns to listen kindly and answers gently:

"She says that it *must* be that God will not let them cry 'How long? how long?' forever. And that they are weary, weary, *weary* with waiting!"

"Will the Temple be restored in all its beauty and given to the Jews?" is my last question.

"Surely yes—for thus it is written, and God keeps His word."

In one of the huts a woman is dying of consumption, a huddle of rags all that shields her worn body from the damp earth. The missionaries have brought her clean clothing and nourishing broth. In another is a baby but three days old, for whom they have flannels and slips and petticoats. With them they leave a leaflet containing a hymn translated into Hebrew.

"A VILLAINOUS-LOOKING TRAMP."

A man, with long curls, and a beard cut in accordance with the prohibition—"Ye shall not round the corners of your heads, neither shalt thou mar the corners of thy beard," clothed in a rusty velveteen gaberdine, a tall cap, edged with a strip of mangy fur upon his head, and a most disreputable bundle in his hand, stops to stare at us as we quit the wretched home of mother and child. He is to American eyes a villainous looking tramp, but Miss Robertson touches me apprehensively:

"He is one of their priests! and will, I am sure, question the poor woman sharply as to what I said to her. He will certainly take the leaflet away from her, should he see it."

Glancing over our shoulders we see him enter the hovel, no doubt with the intention she has indicated.

The good done in this unpromising field by these devoted women is incalculable by any system of human statistics. Walking meekly and unobtrusively in the footsteps of the Master, in sight of the hill upon which He died, they have but one rule of action. "Whatsoever thy hand findeth to do, do it with thy might," when the doing is to succor the oppressed, feed the hungry and minister to the sick. Their parish, their vineyard, their world is the scene before us; their recompense will be given in the day when the Lord of those servants shall come and reckon with them.

There lies open before me, as I write, a printed official report signed by Dr. Selah Merrill, then consul at Jerusalem, of the present condition of the Jews in Palestine, from which I am permitted to glean certain facts, regretting, in condensing the story, that I have not room to copy it in full.

According to this able archæologist and historian, in 1882–83 there was a sudden influx of foreign Jews into this country, so large that, as many readers may recollect, the attention of the Christian world was attracted to what might be the fulfilment of prophecies pointing to the literal return of the scattered tribes to the former home of the race. Dr. Merrill accounts for the movement by the railroad "boom" resulting in the construction of the line from Jaffa to Jerusalem. Many came to look over the ground and to confirm or dissipate the belief that money could be made by the purchase and sale of real estate. Much land exchanged hands during this period of excitement. Of the multitude of Israelites who then visited the Holy Land, a fair proportion remained. Dr. Merrill estimates the number here now at from forty-two to forty-three thousand.

In July, 1891, the Turkish Government forbade the immigration of Russian Jews into Palestine, and land went down one-third in price. The late consul adds that the immigrants are almost entirely of the lower and poorer classes. Well-to-do Jews prefer to live in rich towns, seeking centres of trade. As a race, they are notoriously non-agricultural. Of four hundred and thirty-nine families belonging to the thirteen colonies of Jews established at a comparatively recent date in the Holy Land, two hundred and fifty-five are beneficiaries of the Rothschilds, and, practically, semi-paupers. Rothschild has built model lodging-houses overlooking the Valley of Jehoshaphat and the Pool and Village of Siloam, and besides giving them house-rent and paying water-rates and synagogue-tithes, allows each person a fixed sum per month for maintenance. The like provisions are made with regard to the farm-houses erected upon tracts of arable land in various parts of Palestine, where it is alleged (although on this head I have no data from Dr. Merrill to work upon) that the colonists hire the neighboring fellaheen to do the work at extremely low prices, and do not themselves put the hand io the plough.

"In a word"—thus the ex-consul sums up the case—"Palestine is not ready for the Jews, and the Jews are not ready for Palestine."

CHAPTER XXXIII.

THE CHURCH OF THE HOLY SEPULCHRE.

THE De Credo blood is in fullest flow in the veins of the woman we know as Mrs. Sharpe when the Church of the Holy Sepulchre is spoken of. She actually wept yesterday in confiding to me the pain she feels at hearing so many wise and good people express doubts as to the authenticity of the Church's traditions with regard to it.

"As I said to-day to *my* doctor, *some* regard should be paid to what the church has held for *all* these centuries. It is *bewildering* to listen to his talk of the improbability of *this* and the impossibility of *that* having happened there. What do *I* care about the discovery of a new old wall of the city, which proves that the crucifixion could *not* have taken place where the saints of all ages have believed that it did? And think what a *crash* to the faith of Catholic and Greek Christians if the public *should* come to doubt all the lovely things we are told while in the *darling* old Church! The free-thinking of this so-called Christian age is enough to *curdle* the blood in a pious heart. For *my* part, I agree with *dear* old Bishop Cheeseman, who urges *how* much better it is to cling to what is sanctioned by the belief of *centuries* than to lend ear to a theory not yet *fifty* years old. As to archæology and explorations and excavations and such modern innovations, I have *no* patience with them. They are so many forms of unbelief—downright *Sadduceeism*, I call them."

The ill-matched pair, have by now, ceased to amuse us, and with inward groanings of spirit we see the approach of the husband while we are standing over the flat stone let into the quadrangle just without the entrance of the church. The symbol known as "the Jerusalem Cross," said to have been used as the badge of the Crusaders by the order of Godfrey of Bouillon, is cut deep into the gray slab; all that was mortal of the gallant warrior, made by the decree of his peers, King of Jerusalem, lies under it. We have our own and especial reasons for reverencing the memory of the Christian hero, reasons that have drawn us to the place more strongly than churchly legends, and we are disposed to stiffen up at the prospective intrusion upon our musings.

To our surprise, the usually rampant doubter touches his clerical broad brim in respectful silence, and stands beside us, looking down upon stone and sunken cross, until I am moved to address him:

"I hope that Mrs. Sharpe is well to-day?"

RUINS OF THE HOSPITAL OF ST. JOHN, JERUSALEM—CRUSADER WORK.

"Quite well, thank you," still with the gentle gravity that is to us uncharacteristic. "She is in there"—ducking his red head sideways at the church—and, after a pause—"Praying in the 'Chapel of the Finding of the Cross.'"

It is an awkward moment, no suitable comment occurring to either of the auditors. He resumes, presently, as gently as before:

"I need not say that all the monkish superstitions that bring large revenues to this church are indescribably abhorrent to me. But having expressed this, and more than once, to my excellent wife, and failing to bring her to my way of thinking, I cannot pursue the subject. I can ridicule old wives' fables touching Abraham, Isaac and Jacob, Peter, James and John. Perhaps I lose my temper when the fable is too grotesque. When the subject thus touched has to do with the death and burial of our Lord and Saviour, Jesus Christ"—uncovering his head in pronouncing the words—"I am dumb. Either the imposture is too blasphemous to be dealt with by human speech, or the faith of those who believe this to be the scene of the events I have referred to is too holy to be spoken of lightly. When Mrs. Sharpe told me of her desire to spend an hour in devotion in the chapel below, I raised no objection. I simply answered, 'Very well, my dear. You will find me outside when you are ready.' I more than suspect"—a queer carroty flush mantling the Scotch cheek-bones—"that the sweet soul is now engaged in prayer that my eyes may be opened to see the truth, as it is apparent to her."

With a deepening of respect for a good, if testy, man, and more real liking than we had imagined we could ever feel for him in any circumstances, we enter the ancient edifice.

Those indefatigable church-builders, the Crusaders, remodeled it in the twelfth century, but there was a sanctuary of some kind here in the fourth, a church raised by Constantine in commemoration of the discovery of his mother Helena of what she assumed was the True Cross. As the house in which believers have worshiped for almost fifteen hundred years, it merits reverential mention. We try to keep this in mind and the lesson learned from the usually hypercritical divine just now, as we are arrested every few paces to note this or that holy place.

The stone upon which the Saviour's body lay when anointed for the tomb is near the spot where the Maries stood while men performed the last, sad office, and further away, right under the immense dome of the church, is the so-called Holy Sepulchre. The chapel covering it is tawdry with red marble, gilding and poor paintings. Ever-burning lamps swing from cornice and pillar. We slip off our rubber shoes before we are permitted to enter, lest common soil be carried into the sacred place. In the ante-room to the sepulchre we are shown a rough stone, said to be a piece of that rolled from the door of the tomb by the angels; the place where they stood at the disciples' visit is also pointed out. Stooping low, we pass into a small recess—it cannot be called a room—all ablaze with red, yellow and green lights. There must be between forty and fifty of these lamps, with shades

CHURCH OF HOLY SEPULCHRE, COURT-YARD AND ENTRANCE, SHOWING TOMB OF GODFREY OF BOUILLON.

of different colors, illumining a marble altar, six feet long, three wide, and two high, set out with the customary altar-furniture, gold vessels, artificial flowers and lace-trimmed altar-cloths.

(298) INTERIOR OF CHURCH OF HOLY SEPULCHRE, JERUSALEM.

"This," pronounces David in subdued accents, "is said by tradition to be the tomb of Christ, our Lord."

We linger before it for a respectful minute, and at our movement to retire he continues in the same key, "You will please back out!"

His innocent employment of the undignified phrase does not provoke a smile, but it heightens the incongruity of which we have been disagreeably conscious from the instant of our entrance into the chapel. The necessity of stooping as we retreat backwards, and the dragoman's care of my head lest I should strike it against the top of the low doorway, are a further strain upon the grave decorum we would maintain out of respect to the name, if not the fact of the holy spot.

Outside the Chapel of the Angels we are stopped to look at the "Fire-hole." Perhaps we have heard of it before, but as we listen now the singular tale seems new and incredible on the verge of the twentieth century. A part of the Easter ceremonies of the Greek Church, four hundred years ago—how much earlier we do not know—was the descent of a dove upon the Chapel of the Holy Sepulchre, after which the patriarch, waiting in the chamber of the tomb and invisible to the crowd without, passed a lighted torch through the fire-hole to a priest on the other side. His first care was to light a candle to be sent by a swift horseman to the Church in Bethlehem, and then the multitude pressed upon him, every one with a taper to be ignited at the sacred torch.

"And this is really done still!" we ejaculate.

"The dove does not appear, but the holy fire descends at the Greek Easter, every year."

"And people still believe that the fire comes down from heaven?"

An English-speaking monk, in passing, catches the words.

"And why not?" he interrogates drily, rather than fiercely. "All things are possible with God."

Discussion in the circumstances would be the height of indiscretion. Again taking a hint from Dr. Sharpe's latest lesson, we pass on toward the stairs conducting to the subterranean Chapel of St. Helena. On the way, our notice is called to a dumpy column rising from the marble pavement. I say "column" for want of a fitter word, but it looks more like a raised register or radiator than a monument, and is said to cover the exact centre of the earth.

"Reference is made by those who believe this to Ezekiel, fifth chapter and fifth verse" (David is conscientiously prudent here). "'This is Jerusalem; I have set it in the midst of the nations and the countries that are round about her.'"

"That says nothing of this particular spot. There was nothing here then, to designate it, that we have ever heard of."

THE (ALLEGED) TOMB OF CHRIST IN CHURCH OF HOLY SEPULCHRE.

"Quite so, sir. This is also believed to mark the hole out of which the Almighty took the earth for making Adam."

"What do you think of that story?"

"In my opinion, Adam was never here, sir; but I am not a scholar."

Alcides, bold in the spirit of advanced Young America, does "not take much stock in Constantine. He was probably a charlatan in religion, as in statecraft. If he ever had the vision of the Cross in the sky and the motto, '*In hoc vinces*,' he turned it cleverly to his own account," etc.

I hearken to the fulmination, seated upon a narrow marble bench let into the wall near the top of the steps leading to the Chapel of the Findings of the Cross. By leaning upon the sill of a small square window at a convenient distance above the bench I can look down into a dim cellar, like a tank, badly lighted by two candles set upon what I presently make out to be an altar. Somewhere down there Mrs. Sharpe is kneeling and wrestling in prayer for her husband's soul. Here, if we heed tradition, sat the aged Empress Helena in the year 326 A. D., and watched the excavations going on at her order, in quest of the True Cross. She had dreamed before setting out upon her pilgrimage to Jerusalem, where this precious relic would be found.

A MEANER FIGURE.

I interrupt the recital here:

"She could hardly have sat upon this bench and looked through this window, for they are a part of the church which was built to commemorate 'the Invention'—in ecclesiastical phraseology—'of this very Cross.'"

"Invention isn't bad in the connection!" Alcides slips in an "aside."

The narration proceeds. Three crosses were dug out of the hole down into which I am gazing, and the Empress, the shrewd mother of a shrewder son, bethought herself of a test that should reveal the right relic. A Christian woman of rank lay on her deathbed in Jerusalem. The three exhumed crosses were borne into her room, and by touching one of them she was healed. This is the one and only ground, so far as we have been able to discover, for the belief that the Church of the Holy Sepulchre rises above the place of crucifixion, burial and ascension of our Lord, unless we are to admit the evidence of a church historian of that day who ascribes the building of the church to Constantine, who had of himself discovered the site of the Sepulchre.

When we consider the improbability that the crosses upon which three Jewish peasants, criminals in the eye of the Roman authorities, would be preserved in any way after they had served their ignominious end; the greater improbability that they were buried near the sepulchre of the Nazarene; when we take into account the three centuries during which they thus lay concealed from the knowledge of mankind, and then, the manner of the disinterment, one grows thoughtful and distrustful. The students of God's Word, who in trying to identify any locality in or near Jerusalem with the scene of our Lord's crucifixion "without the gate," have more than one discrepancy to confirm them in the belief that this venerable church does not cover the spot.

"THERE ARE A GREAT MANY OF THEM."

Vespers are singing or intoning in the Chapel of St. Helena while we loiter in this corner. Much visiting of churches has accustomed our senses, and not unpleasantly, to the throbbing echoes awakened in aisle and dome by responsive chanting and the organ accompaniment. We rather like the smell of good incense, and the æsthetic sense is gratified by the "dim religious light" produced by the blending of lamp-rays and the faint daylight that finds its way through the stained glass windows.

Just at present we are in no mood for enjoying æsthetic effects. Even Mrs. Sharpe, as Miss De Credo, must have supped her fill of superstition and churchly tradition in her frequent visits to the sanctuary reared by the Crusaders. *They* believed, to the bloody death, in the authenticity of the story that makes the gorgeous altar in the low-ceiled chamber over yonder the tomb from which our Lord arose upon that first Easter morning. From the depths of aching hearts we wish that we could credit it, and—say one-tenth of—the other tales poured into our ears with volubility acquired by continual practice. We look the sacristan who has us in tow steadfastly in the eyes for sign of faltering or embarrassment, as he runs off legend after legend as an auctioneer extols his wares.

Golgotha has three chapels attached to it, each erected by a different sect of Christians. It is even harder to believe in than in the sepulchre, and is yet more effectually disguised by precious metals and marbles. The square opening above the socket in the rock in which the foot of the Cross is reputed to have been sunk, is lined and bound with silver. Scarcely six feet away, is what looks like a ventilator of open brass-work that—upon payment of a fee—is slid aside to show a

A BEGGAR'S NOON SIESTA.

fissure in a rock, said to have been made by the earthquake that rent the veil of the temple during the crucifixion. In close succession are exhibited with business-like promptness the spot where Mary Magdalene fell upon her knees, exclaiming "Rabboni!" the grave of Joseph of Arimathea; the prison in which our Lord was detained until the Sanhedrim could be collected; the tomb of Nicodemus, and of James the brother of John and son of Zebedee, who was killed by the sword in Herod's persecution of the early church; the Pillar of Scourging; the place where the Roman soldiery parted the raiment among them, casting lots for the vesture; the boughs in which the ram was entangled as a substitute for

Isaac, and the altar on which Abraham would have sacrificed his son ; the tomb of Adam in whose dust we are told the Cross was set up—and so many other notable places that I have not the heart to transcribe a list which could not but disgust the sensible reader, and pain the devout.

We are driven, whether we will or not, to recollect at every new exhibition, Dr. Sharpe's allusion to the revenue drawn from this marvelous—I had almost written monstrous—collection of sacred *curios*, and to wonder, only, at the economic instinct that has gathered so many under the one domed roof of the Holy Sepulchre. Nor is it possible, being human, and readers of the Bible, that we should fail to remind ourselves and one another, in connection with the venders of rosaries, charms and photographs without the church, and of painted candles within; of the riot raised by the craftsmen of Ephesus when the trade in silver shrines for Diana was threatened by Paul's preaching.

"I DON'T SEE WHY NOT."

The ablest scholars who have studied and written upon sacred history, agree in declaring that this cannot be the true Calvary ; that the site of the Church of the Holy Sepulchre never was within the city walls, and corresponds in no particular with the description of the holy spot given by the Evangelists. The fact will never be admitted by those whose interest it is to encourage pilgrimages to the Mecca of Christendom.

It is like a breath of purer, honester air when David again takes us in hand and conducts us into a side-room to see the sword and Jerusalem cross-badge worn by Godfrey of Bouillon. They are kept in a locked coffer and shown, upon the payment of a trifling sum, to the sacristan. The Syrian dragoman lifts the great weapon carefully, and lays it in Alcides's hand with :

"There, sir, is the sword of your favorite hero. He must have been a great man in more ways than one, if he used that in battle."

The badge is several inches long, and both it and the sword are evidently extremely old. Our interest in them being merely a matter of hero-worship, we do not tread upon, to us, debatable ground in choosing to believe them genuine relics of the stout-hearted, stainless knight of St. John.

WAYSIDE BEGGARS.

At the head of the steps leading into the dismal Chapel of the Invention of the Cross, sits an old man with a noble head and a white beard, such as we see in pictures of Abraham. He is turbaned, and wears the brown-and-white abieh of the fellaheen. He raises his face at sound of our footsteps, and we see that he is blind.

"For the love of God and for holy charity!" he quavers, holding out a tin cup.

"A professional beggar," I remark. "He might be King Lear, or Belisarius, or Homer. Get his story, if you can,—please?"

He tells it readily, but not officiously. According to it he is a farmer from Ophrah.

"If the Captain will look up 1 Samuel, xiii. 17, when he gets back to the hotel," murmurs David aside and parenthetically.

While on a visit to Jerusalem twelve years ago, he went to sleep one night, well, and, awakening next morning could not open his eyes, "for a great swelling which made his head so large"—describing an area about it with his hand. He went to a physician who put something into his eyes that burned like fire, or boiling water, and from that time he has been totally blind, unable to tell day from night. A blind man cannot work in the fields or tend cattle, or thresh grain, so he came to this city and makes a living by begging. He has children who give him a home where he may sleep at night—but they are poor—very poor—and he will not be a burden to them. Being a Christian, he is allowed to sit here and ask alms of those who come to pray, or to see the church. Sometimes, in the season of visitors or at Christmas and Easter, he makes as much as four piastres a day (fifteen cents of our money), sometimes he makes but two piastres, sometimes but half a piastre,—sometimes but eight paras—(the lowest in value of Syrian coins, being many times less than a cent).

David is still translating when Mrs. Sharpe comes up the steps, rising oddly out of the darkness, first, her white face, then her hands, becoming visible against the black background and her black dress. She stops to listen, and we see that her eyes are large with tears, her face pure and solemn, for all the babyish roundness and softness it will never outgrow.

"And what do you think, *then*, of a heavenly Father who lets you go to bed hungry?" she interpolates as the last pitiful sum is named.

"I go to sleep and trust Him for to-morrow," the man makes answer in simplicity that sounds sincere.

"Tell him that is right," Mrs. Sharpe instructed David—"and never, *never* to doubt Him. Tell him too; how sorry *I* am that he must sit here all day long in the darkness and where it is so damp, and ask him to pray for me to-night."

She has dropped two francs into the tin cup, and hurries away to avoid his thanks.

"I believe that is a good woman," utters David, peering into the cup where the silver shines bright upon a layer of copper coins. Then, to the beggar,—"You have there two francs. When you buy your supper, see that you do not pass them off for pennies. And don't forget to pray for the lady."

A second beggar sits at the head of still another staircase, and here, too, the sunshine never falls. He is a meaner figure than that we have just left, but is made picturesque by a pretty little girl, not more than eight years of age, who leans against his knee and regards us with big, solemnly pathetic eyes. They have been here all day and come every day in the week, the child leading him from his home outside of the city walls to his place upon the steps about eleven o'clock in the day, few tourists visiting the church before that hour. The child looks like a plant that has grown in a cellar, pale and slight to fragility, so poor of blood that she shivers all over now and then, under her thin cotton gown. Her brown feet are bare, and her paleness is made more striking by a black shawl worn over her head. Her father was blinded by a sunstroke while working in the fields before she was born, and has no business but beggary.

"I have been told that every professional beggar in Jerusalem would be cared for by the convents and other religious organizations if he would let them help him," I say to David when we are out of the church and climbing Christian Street.

"There are a great many of them, madam, and as they must be supported by charity, having no work to do, and knowing no trade of any sort, they are freer to come and go when they beg for themselves from anybody who happens along, than if they had to obey rules and accommodate themselves to certain hours, and all that. To them, their way of living is as respectable as to ask alms of a religious society. They may be wrong, but that is the way they feel about it."

The idea is novel—and there may be something in it—in Jerusalem. The poverty of means and of resources prevailing among the lower classes here is patent and pitiable. The afternoon is fine and cold, and we walk from one side of the city to the other before returning to the hotel—not an arduous undertaking, the entire circumference of the walls being less than three miles. We are hurrying somewhat, in order to get back in season for dinner when, for the second time to-day, we encounter Dr. and Mrs. Sharpe. Whereas we were glad of the opportunity afforded by a former meeting to modify a harsh judgment, we wish now that we had taken another turning of the narrow thoroughfare in which we find them. For the husband's hair bristles pugnaciously, and the wife has on her most amiably-obstinate expression.

"I don't see why *not*, doctor dear," we hear in nearing them, "What more natural than that He *might* have pointed to that very stone in speaking the words, and devout men would be *sure* to treasure it piously afterward."

"Of all the sacrilegious enormities,—" splutters her spouse, and we quicken our speed to get out of hearing.

The cause of dispute is known to us, having been designated by Dr. Merrill in one of our earlier walks about Jerusalem. It is a time-stained stone built into the wall of a filthy cross-street, a round, common-looking fragment, with a hole in it, an aperture enlarged and discolored by the kisses of devotees. The relic is believed by others besides Mrs. Sharpe to be "the stone that would have cried out," had "the whole multitude of the disciples rejoicing and praising God with a loud voice," have held their peace in the Master's triumphal progress into the City.

That the appeal of the Pharisees to Him to silence the acclamations, and the Lord's reply are plainly said to have been spoken at the descent of the Mount of Olives, avails nothing with those who make the word of God of none effect through their traditions.

Can human effrontery and the credulity of superstition go further than this?

THE FIELD OF BLOOD, JERUSALEM.

CHAPTER XXXIV.

TO MAR SABA.

THE palankeen has been made ready instead of a horse, for my use in the journey from Jerusalem to Mar Saba, said to be the oldest convent in the world. I acquiesce the more willingly in the arrangement that the memory of a succession of "Jerusalem ponies" does not incline me toward the experiment of undertaking a long ride upon what I heard an Englishman describe in a Christmas talk in Bishop Gobat's school on Mount Zion, as "that most unhasty beast, the ass." I do not take kindly to the donkey nor, to judge from his behavior when honored by bearing my weight, does he to me.

Wise David demurs when a horse, a slow and amiable creature and gentle of motion, is proposed. The way is steep and rough, he represents, and much time would be lost by a certain traveler's habit of alighting to walk up or down particularly dizzy heights. The mules and donkeys are as sure-footed as cats from long training, and madame has faith in her muleteers. Madame, growing indolent, and maybe a trifle weary in nearing the end of her many and varied journeyings, seconds the motion.

We leave the Jaffa gate at noon of a lovely day that has in it a promise of spring to the bare earth. Already pale purple and bright yellow blossoms show bare heads above the withered turf on sunny terraces, and in sheltered nooks looking southward. As we wind below the bluffs beyond Siloam, built up with the mean huts of the leper settlement, we see upon the right, the Potter's Field, or Aceldama, in which, it is said, Judas hanged himself. A tree that must be nearly, if not quite, fifty years of age, is pointed out as that which refused to bear the traitor's weight. By now, we have learned to laugh at such solemn absurdities, and, a more difficult undertaking, not to let a manifest impossibility blind us to what may be true and what is altogether reasonable.

A little further on we meet a party of a dozen women laden with enormous bundles of the dried furze or low thorny growth which serves them for kindling wood, and is, in mild weather, the only fuel used by many of the poorer peasantry. They resemble nothing else so much as ants plodding along under burdens many times larger than their own bodies, but this company is unusually merry, talking, laughing and shouting gayly to each other as they take the side of the road to give us the middle. One is really very pretty, bright-eyed, plump and light of foot, although she carries, besides her great bunch of prickly stuff, a baby slung between her shoulders in such a bag as we bought from "Martha of Bethany." She does not stoop under the load, and glances laughingly in at the window of the palankeen from the stony bank that brings her face on a level with mine, calling out something, to which David replies good-humoredly.

"What did she say to me?" I inquire in natural curiosity.

"That you are blessed among women," is the translation given to me, but presently Massoud appears at my side, and Alcides supplies the rest of the young mother's salutation.

"She asked if you wouldn't get down and give *her* a ride. Human nature on one side of the world doesn't differ much from human nature on the other."

"HER PEOPLE."

I comfort myself by the belief that the brown beauty's challenge was mere banter. The tone was too cheery, and her smile too free for envy.

The route is all new to us, and after passing the fields outlying the city of Jerusalem, we decide that we have seen no more desolate and forbidding country. Except for a few scattered patches of vegetables cultivated in the close neighborhood of what are water-courses for a few months of the year, and an occasional olive-grove, also in the low grounds, not a glimpse of green blesses eyes pained by dwelling upon gray rocks and livid hillsides. For a while, our course lies in the pebbly bed of the Brook Kidron, now as dry as dust, and glaringly white. Then we begin to climb by a narrow, twisting path, where the horses walk in single file and the palankeen, scraping naked rocks upon one side, overlooks, upon the other, sheer precipices from fifty to two hundred feet deep. Looking forward, I am often bewildered to guess where, amid the heaps of stones and the criss-crossing of projections that seem to close up the route, my faithful mules can make their way, but they do not slip once, even when their small, nimble hoofs displace loose pebbles and larger stones, and send them clattering into lower ravines.

There is so little to amuse me and we pass so few notable points, ("places" there are none) that I fall to watching a woman with a mass of something green upon her head, walking on the hillside in a straight line in the same direction with ourselves. As the route she holds with something of the crow's instinct in determining the shortest distance between two given places, brings her near enough for me to see her more plainly, I perceive that she wears a dark blue cotton tunic, reaching to her ankles, which, like Maud Muller's, are "bare and brown" as well as her feet. The tunic, or underdress, has wide flowing sleeves; over it is a long sleeveless jacket of red woolen stuff; an embroidered and fringed veil, once white, now dirty, drapes her head, and one end is bound about her burden which is composed of the refuse leaves of cabbage and cauliflower. Without casting a look at our party, she holds on her way, never quickening or slackening her fleet walk, down hills and across torrent beds, springing from crag to crag like a chamois, until I see her near a gap in a ledge hardly wide enough, apparently, to afford a foothold to a mountain-goat. The shelf of rock follows the face of a bare granite shoulder of the mountain towering above us until the whole valley is darkened by the shadow. The gap has been made by a rush of winter rain, or by a landslide, and looks to be about four feet wide. As the girl approaches it, she puts up one hand to steady the load upon her head, and, without stay or falter, leaps across, landing erect upon the farther side, then passes on as fleetly as before. We have never seen another woman walk so gracefully and fast, and call upon David to find out who she is and upon what errand she is bound.

The dragoman gallops forward and intercepts her where his practised eye has seen that she must keep our path for a while, and we see them talking together for a hundred yards or so, the girl actually lessening her speed to keep back with Dervish's walk. Next, both have stopped to wait for Serkeese who is trotting along in the gravelly "bottom," atop of hampers and luncheon-tent. When palankeen and Massoud come up with them, the woman is still walking beside the mounted dragoman, devouring a loaf of bread drawn from Serkeese's stores, as incurious and impassive as ever. David falls back to my side and narrates:

The young woman went to Jerusalem this morning to buy some cloth in the name of her tribe, only to be refused credit by the merchant to whom she was sent. She had no money to buy food, and picked up the refuse greens in the market-place. Breakfastless and luncheonless, she set her face homeward and has now walked more than fifteen miles since daybreak. While he speaks, she looks back to wave her hand, and disappears around a hill. From the parting of our ways, we, by-and-by, see her swiftly climbing the long breast of the hill, near the summit of which is a cluster of black tents, with moving figures before them.

"I hope she will not get a beating from her people for her bad luck in not buying the cloth," says David, solicitously—and when we admire her strength and staying power—"oh, she is used to walking all day, and to fasting."

Wilder and gloomier grows the way, but now we are in a veritable road, winding up and along the heights, a low parapet of loosely-laid stones guarding

"THE ANCIENT EDIFICE."

it upon the outer edge. The sun still sleeps upon the heads of the gray and reddish mountains, but in the depths of the defiles night is settling. This is the wilderness in which John the Baptist spent his novitiate, these are the deserts in which he "grew and waxed strong in spirit, till the day of his showing unto Israel."

"And the same John had his raiment of camel's hair, and a leathern girdle about his loins, and his meat was locusts and wild honey."

The rude figure thus portrayed harmonizes perfectly with the naked gorges where not a leaf of herbage or blade of grass clings to sides honey-combed with caves. These natural dens and grottos were, before the forerunner of our Lord sought their awful solitudes, the resort of fanatics and world-weary hermits. In the early ages of the Christian church, literally thousands of refugees from persecution, and men sickened out by the corruptions of society, fled to the "Wilderness of Judea," and dug cells in the cliffs in which to hide themselves for prayer and sacred meditations and fastings innumerable. About 460 A. D., the leading spirit of the strange colony, a Greek hermit by the name of Sabas, established at the head of the gorge the monastery that bears his name. In time, repeated

TOWER OF JUSTINIAN AT MAR SABA.

attacks from predatory bands, and, early in the seventh century, a terrible invasion of the Persians in which 3000 monks and anchorites are said to have been massacred,—made it necessary to fortify the retreat, and it became almost impregnable.

The sun is near the horizon as we gain a plateau behind the ancient edifice which hangs dizzily over the "Valley of Fire," four hundred feet below the foundations laid in the solid rock. We do not ask for admittance. By an inviolable law of the establishment, no woman can pass the outermost gate, and, while, in a detached tower, removed by some fifty feet from the forbidden precincts, shelter and food would be given me upon application, I prefer the luxurious inde pendence of our camp. The evening is cloudless and bland, and the moon in her second quarter lights up the tall tower erected by Justinian, the grim walls and the belfry, from which the hours chime out sweet and startlingly clear in the rarefied air. Seated, as usual, in the tent-door, we hearken to stories of the stringent austerities practiced by the holy brotherhood. They eat no meat, and are allowed one egg apiece on Sunday, black bread, vegetables, a scanty allowance of fruit and sour wine forming their diet the year around. There are seven services in the twenty-four hours, the first long before daylight, and, the vows once taken, the monks seldom quit the convent upon any errand.

Yet the loftiest mountain of the range separating them from the world of active labor is believed to be the Hill of the Scapegoat, down which the hapless animal, bearing upon his head the iniquities of the people, was thrown yearly by the hand of the "fit man" who had led him into the wilderness.

I think of the ceremony and its significance as, awakened in the dense darkness of a winter morning at four o'clock by the vibrant call of the bells to prayer, I put out my hand for the extra blanket laid across the foot of my bed, and, nestling down in the warm comfortableness of my nest, picture the shivering monks kneeling for two long hours upon the stone floor intoning prayers to Him who, like as a Father pitieth His children, and has given them all things richly to enjoy.

THE FACE OF THE CLIFF AT MAR-SABA.

CHAPTER XXXV.

AT MAR SABA.

THERE is a profundity of silence that hinders sleep. I have never appreciated the fact more sensibly than in the early hours of the new day we greet from the bald mountain-top on which our camp is pitched at Mar Saba. The world is in a dead swoon. From the time the bells chime out for the first service, until I hear the shuffle of the muleteers' bare feet upon the rock as they feed their animals, and the subdued clatter of pan and

"IN SILENCE AS SULLEN."

dish in the kitchen-tent testify that John is getting ready to feed *us*, the beating of my own heart is absolutely the only sound I hear.

We have finished breakfast before we have any tokens that there are other living things besides ourselves in the vast solitude. Then the interruption is sudden and peculiar. Alcides is filling his fountain-pen, and I am unscrewing mine to see if it also needs attention, when right between us fly two birds, tied together by a leash that, getting entangled about Alcides's legs, checks their flight. Around the corner of the nearest tent comes David in pursuit, and at his heels the more deliberate figure of a Bedouin, who looks on in apparent unconcern while the pretty creatures—a species of orange-and-black grackle unfamiliar to us—are captured.

"HELD TENDERLY BETWEEN HIS STRONG HANDS."

David has had an order from Lord Somebody, an English or Scotch nobleman, who is a zealous ornithologist, for five hundred of these birds. As they are to be found nowhere except in this neighborhood, and about the southern shores of the Dead Sea, it is necessary to depute a native of the region to procure them, and our Bedouin friend has the commission.

"He is a great cheat," subjoins the dragoman. "He wants me to pay him four shillings the pair, when he has snared them without difficulty in the valley down there at the bottom of the pass! They are very tame, for the monks feed them there three times a day. He is a rascal who has no conscience whatever."

While he grumbles, the birds, held tenderly between his stout hands, have fastened upon his thumbs with all the strength of their little beaks. The feeding of them within the convent walls is one of the few human enjoyments vouchsafed to the brothers of Mar Saba. Other wild things—jackals, foxes and wolves—make friends with the recluses and have their stated times of feeding, regulated—as we are told, and it is possible with truth—by the chiming of the bells as they mark the hours of service. One's heart contracts with an odd physical pain in hearing of this phase of the starved, narrowed, belittling round of existence decreed by the leaders of these men's souls.

SOME OF OUR VISITORS AT MAR SABA.

The current of compassion is somewhat changed by the appearance of a holy brother in the costume of the order who leaves the pale of sanctity, bearing to the pale-faced strangers, not the blessing and good-will of the community, but rosaries, wooden forks and spoons, charms, walking-sticks and other traps for the coin of unwary heretics. He spreads his wares in sullen silence upon the rock, and in silence as sullen, stands looking down at them. The Bedouin "cheat" has drawn forth a long pipe from the folds of his abieh and smokes in tranquil contentment, sure of getting at least one-third of what he has asked for the half-domesticated grackles; his bare-legged son gapes at the display of "curios" I am turning over with uncovetous fingers, while the kodak gets the picture of a dark-visaged, bad-eyed fellow whom one would not choose to meet in one of the lonely ravines intersecting the Valley of Fire. In fact, as the dragoman informs us, this famous monastery is in our degenerate day a sort of penal settlement for monastic culprits. A member of the present fraternity is a notorious murderer from Port Saïd, his residence here being equivalent to imprisonment for life. Our merchant makes no more effort to sell his wares than if he were a young saleslady, brave in frizzettes and rhinestones, behind the counter of a New York emporium. If we want anything, we can take it; if not, leave it. His sulkiness increases rather than diminishes at David's courteous petition to be allowed, upon payment of a gratuity, to conduct Alcides into the convent. Gathering up his wares, the

CONVENT SEEN FROM THE TABLE ROCK.

brother restores them to his basket and stalks on ahead of us to the little table-land under the walls of the detached tower where I am to be left.

"If he is a specimen article, I do not envy you your visit," I say philosophically to the favored pair, and settle myself for half an hour with my note-book.

A folded rug makes the rock passably comfortable as a seat; the air is still and bracing, the sunshine delightful. I have written fast and satisfactorily for perhaps twenty minutes when a shadow strikes the page, and I raise my eyes. Two women, one middle-aged, one young, both clad in the dark blue gowns and veils of the Bedouins, have arisen as out of the rock, and are staring placidly at me. Behind the older woman a boy of five, or thereabouts, is peeping around her skirts. To test their mood and manners, I write on in serious disregard of their presence. The mother sinks to the ground and watches my fingers and pen, intent and mute. The girl lifts my glove and veil from the rock where I have laid them and examines them closely, and as I give no sign of noticing her occupation, goes on to finger the trimming upon my skirt, calling her mother's attention to it in childish glee, then passing her fingers lightly up the cloth of the garment until she reaches my sleeves, picks at the braiding to see if it will come off. The whole proceeding is as inoffensive as if she were a baby, but I do object when she lays the tip of her finger upon my pen. At my gesture of disapproval the mother chides the meddler sharply, and pointing to the convent, evidently

GRACKLES FEEDING IN COURT-YARD OF CONVENT.

tells her that I am writing of, or drawing it. Thenceforward, two pairs of wondering eyes follow the flowing ink, looking from the page to the building and back again until the exhibition is too much for my gravity. I sheathe the pen, close the book and laugh outright in their faces. After a slight start of surprise, they join in the merriment and fall to asking questions, not one of which I comprehend.

I wear upon the front of my corsage a brilliant carnation, brought from Jerusalem, and presently drop it into the girl's hand. She exclaims with delight, smells it, shows it proudly to her mother and finally hides it carefully in the bosom of her gown. Whatever glory of blossoming spring may bring to these sad-colored wilds, it is plain that a flower in winter is a phenomenon to the untutored daughter of the desert. She would be pretty but for the blue tattooing disfiguring her face. A line of polka spots runs clear across her forehead; another row defines the lower lip and three are "powdered" irregularly upon one cheek; a tattooed bracelet of Grecian pattern encircles each wrist and a square of the same design is upon the back of her right hand. Seeing me look at them, she laughs, pulls a long, cruel-looking needle from the front of her mother's gown, and goes on to show me how the marks have been made, and that her mother is the artist. When I shrink and signify that the operation must be painful and disagreeable, she laughs again—the infantine gurgle one never hears in our country from a child over ten years old—and pulls up her loose sleeve to display a more elaborate pattern sprawling all the way to her elbow, the upper part still raw from the needle.

"ARRESTED UPON THE HALF-STEP."

Two ragged children have joined the group, and their inquisitive forefingers are unpleasantly familiar with my dress and portfolio. Twice the mother, at my appeal, boxes their ears, but without outcry or any sign that the blow is unwelcome, they return to the charge until I get up, collect my belongings and walk to the edge of the table-rock almost overhanging the courtyard of the monastery. Two black-robed friars are pacing a short balcony in the sunshine like bears in a cage; in the stillness I can hear the whirr and whiz of a hundred pairs of wings. The grackles are feeding upon the crumbs and grains flung from the front of the convent into the defile, and strewed upon the pavement of the inner courts.

I have not shaken off my juvenile tormentors. One tugs at my skirt, another lifts the flap of the writing-case in my hand, a third, the biggest and dirtiest of the trio, and who rejoices in the only pair of shoes in the party (which, by the way, remind us of the Gibeonites' stratagem of "old shoes, and clouted, upon their feet") pushes impudently to the front and demands "baksheesh." At this opportune moment I espy David and Alcides in the convoy of a lay brother descending an outer flight of stairs in going from one wing of the building to another, and call cheerily to them. The dragoman takes in the situation at a glance and shouts out something to my besetting neighbors that frees me from annoyance. They fall into the background and remain there until the return of my escort.

The girl, who must be about sixteen, withdraws modestly, putting a corner of her coarse linen veil over her mouth at sight of two men, one young and pale-skinned. As I am determined to get a picture of her, all David's tact is brought into play to distract her attention and her mother's suspicion from the tell-tale kodak while this end is secured. She is actually in flight, arrested upon the half-step by the dragoman's call—an untamed creature as timid and wild as one of the hares we frightened out of a hollow yesterday. Her brothers and father are easily persuaded to stand for their portraits, but we see her no more, until we drop down, as it were, upon a group of black tents from a steep ridge two hours later, and recognize her as the centre of a knot of attentive listeners, enchained by the narrative of her morning "outing."

CHAPTER XXXVI.

HEBRON.

I MADE me great works; I builded me houses; I planted me vineyards; I made me gardens and parks, and I planted trees in them of all kinds of fruits; I made me pools of water to water therewith the wood that bringeth forth trees."

So writes the Royal Preacher in the preface to the cry of "Vanity of vanities! all is vanity!"

"So I was great, and increased more than all that were before me in Jerusalem," was no unfounded boast.

The king who projected Solomon's Pools on the road to Hebron was so many generations in advance of his predecessors that, reasoning after the manner of men, we think that he might have excepted this gigantic enterprise from the sweeping condemnation passed upon the rest of his "great works." The three immense tanks, to visit which we have left the direct road to Hebron, are lined with hewn stone, and in the opinion of wiser critics than my unlearnéd self, were constructed by Solomon's workmen to lead the living waters of adjacent springs to his pleasure-gardens, the "orchards of pomegranates, with pleasant fruits, cypress with spikenard and saffron, calamus and cinnamon with all trees of frankincense, myrrh and aloes, with all the chief spices."

The enumeration is plainly incongruous with the ruined reservoirs and the sereness of the surrounding fields at this season. The air "nips shrewdly," despite our heavy furs and the clear shining of the sun. Midway between the highway—the finest in the length and breadth of Palestine—and the Pools, is the "sealéd fountain" of Canticles, a square building covering the source of the waters led by an underground aqueduct to the lower reservoir. Close to the walls enclosing the great cisterns is another small building, with a domed roof, erected above a second well, reached by six or eight steps. The water is very cold, the steps are wet; it makes one shiver to look down into the dark cavity; but half a dozen women are passing down to fill water-skins, and then lugging them up. A water-skin or bottle is the whole hide of a goat tanned inside and out, sewed up lengthwise, and filled through the throat at the well. It is carried by means of two cords tied about the front and back feet, then passed across the forehead of the bearer, a fold of her veil keeping the rope from cutting into the skin. As the skin is always of a goat three-quarters or full-grown, some idea may be formed of the weight of the burden. Two of the women with whom we speak at this well live a mile away, a third at Bethlehem, more than twice as far.

We have regained the road when we catch sight of a man ploughing with a camel, the first instance of the kind that has come within our observation. The slight wooden plough and the undersized man contrast grotesquely with the unwieldy brute, but the incident leads us to note the increasing fertility of the country. The grapes of Hebron are noted for size and flavor and clothe the upper terraces of the hillsides. Lower terraces are clothed with fruit trees, the

"SEALED FOUNTAIN."

fig and mulberry being the most abundant. There is a proportionate improvement in the looks of the husbandman. They are better clad, more alert in movement and really work as if a motive lay back of action. Women are busy in the fields, pulling up dried furze by the roots; donkeys, laden out of sight except for their ambling legs and the tips of their noses, are met in droves; camels, hitching their clumsy bulk along under moving groves of a stouter shrub that perfume the air as they brush us in passing; bearded Moslems with white turbans coiled about shaven heads, and (as we never fail to think, in contemplating the close fit of the head-gear) the few valuable papers they possess in the world snugly stowed away between the inner and outer linings of the turban,—are indices of thrift and a fair degree of prosperity.

TREE OF ABRAHAM, HEBRON, HOLY LAND.

Our guide points to a level stretch a mile away from the road, and we cease to note or consider anything modern.

"Mamre!" he utters, "and with this glass you can make out the old oak, the only tree of the kind in many miles."

Not *the* oak, of course, although we talked last week in Jerusalem with a man who argued earnestly for the possibility of the existence, through thousands of years, of the tree under which Abraham "dwelt in the plain of Mamre and built there an altar unto the Lord." But we gaze in the direction indicated with intense and devout interest. From that plain or plateau Abram sallied forth, at the head of three hundred and eighteen servants born in his house, to overtake and defeat the kings who had taken captive his brother's son; there he had the vision of the smoking furnace and burning lamp passing between the "pieces" laid in order for the sacrifice, and received the exceeding great and precious promise of the son to be born of his old age; there, from the home where life had grown intolerable, Hagar fled into the wilderness to be sent back by the Divine command; there, "the Lord appeared unto Abraham in the plains of Mamre, as he sat in the tent-door in the heat of the day," and upon the next day from the elevated table-land, the anxious patriarch looked toward Sodom and Gomorrah to see "that the smoke of the country went up as the

WOMEN CARRYING WATER-SKINS AT SOLOMON'S POOLS.

PIGSKIN WATER BOTTLE.

smoke of a furnace." The cave purchased from the children of Heth as a burying-place where the dead wife could be buried out of the husband's sight, was in "the field of Machpelah before Mamre. The same is Hebron in the land of Canaan."

Besides the Prince of Wales, his sons, and their attendants, including Dean Stanley, but five or six people, not Moslems, have ever been admitted to the interior of the Mosque of Abraham. Gen. Lew Wallace, by special permit from the Sultan, obtained this privilege, and took with him several particular friends

MAN PLOUGHING WITH CAMEL.

of his own, among them Dr. Selah Merrill, then resident in Jerusalem. Their report was of six mock tombs of Abraham, Isaac and Jacob, Sarah, Rebekah and Leah, each wife lying opposite her husband in the sealed crypt below. Beyond these, the mosque contains little to interest Jew or Christian. "Father Abraham" occupies an exalted place among saints revered by the Moslems, and the jealous hatred of the Jews, never absent from the creed and feelings of the worshiper of Mohammed, is at fever-heat in Hebron. Nowhere else in the Holy Land, or out of it, are they regarded with such intolerant suspicion as in the ancient city in which David reigned over Judah seven years and six months. Hence, the approach of an Israelite to the tomb of the patriarchs is even more abhorrent to "believers" than that of the "Christian dog." On our way to the Mosque we see certain money-changers of the despised race, sitting in the bazar, and learn how cunning is their revenge upon the masters who would drive them clean out of the land if it were possible. Traders and usurers everywhere, they grow rich in the neighborhood of Abraham's tomb by becoming money-lenders and pawn-brokers to the poorer citizens, and especially to the farming peasantry. One, in whose physiognomy the characteristic facial features are conspicuous, is examining a set of silver ornaments belonging to a farmer's wife, her husband standing anxious by her side.

"He will lend her one-tenth of what they are worth," says David, "with the certainty that she can never redeem them. Or, he will buy them, out and out, at one-fifth of their value. I know his reputation. He has no mercy, and indeed it is not strange that he should be hard with these people who despise and insult him."

A flight of broad, low steps conducts the faithful from the street to the main entrance of the mosque. Beyond the fifth step none but a Moslem may go upon penalty of death. We ascend to the forbidden line and look defiantly up, then aver that there is nothing within the walls worth our seeing, and enact the King of France's celebrated retrograde movement, stopping on the way "down again" to peer into a deep crevice between two of the old stones of the wall, into which, under the easy tolerance of the present government, the descendants of Abraham may, when they like, thrust their arms at full length to drop written petitions to their great ancestor. This is the Hebron Wailing Place and their nearest approach to the tomb. The ceremony must be some sort of sad satisfaction to them in their disreputable homelessness in a land once deeded to them by the Judge of the whole earth, for we are told that many avail themselves of the poor right.

A wall of comparatively modern masonry shuts in the mosque from profane eyes. By mounting a heap of rubbish across the street we can see the quadrangular building reared above the cave of Machpelah. The gray walls are of the great stones we have learned to recognize as belonging to the Phœnician period, and identical in finish with those in the base of David's Tower in Jerusalem. The Crusaders consecrated and used the mosque as a church. A writer of the period during which it was occupied by the Christian invaders describes the cave as divided into three chambers, the last containing the six tombs of which I have spoken. He relates, also, that it was the habit of Jews to bring thither in tubs or boxes the bones of their dead to be laid near the sepulchres of the patriarchs. The desire to sleep with one's fathers and the custom of the survivors of carrying out the wish were even then extremely ancient. Witness the oath exacted by

Joseph from the children of Israel that they would carry his bones out of Egypt with them, a promise kept two hundred years afterward.

A narrow alley leading to the mosque is alleged by tradition to be the scene of the murder of Abner by Joab, when "he took him aside in the gate to speak with him quietly and smote him there under the fifth rib that he died." Assassination so dastardly that every reader sympathizes in the bitter outbreak of the nomiinal king then reigning in Hebron; "These men, the sons of Zeruiah, be too hard for me!"

Our next halt is at the "pool of Hebron." It is a large shallow tank, occupying about half as much space as a modern city block of average size, and sur-

THE CITY OF HEBRON, AS IT NOW APPEARS.

"Here David Reigned over Judah."

rounded by a wall of solid masonry. The water is stagnant, and coated with green scum, but women are coming down the steps at one corner with skins to be filled. The water, such at is, is shallow, leaving exposed above it, some twenty feet of wall. We try to guess whereabouts were nailed the feet and hands of Rechab and Baanah his brother, the sons of Rimmon the Beerothite, who, after killing the sleeping Ishbosheth, "took his head and gat them away through the plain all night and brought the head of Ishbosheth unto David to Hebron."

Instead of the expected reward, they were put to death and their hands and feet hung over this pool as a terror to other evil-doers.

Our luncheon is spread in the house of the only missionaries in Hebron, Mr. and Mrs. Murray. They are, moreover, the only Christians in the home of Abraham and David, with the exception of an English family resident in the hospital lately erected here by the Church Missionary Society. Mr. and Mrs. Murray are under the care of no one denomination or society, and lead the simplest, happiest "life of faith" it has ever been my privilege to behold. In one room is collected a class of twenty-two little Moslem girls, who are instructed in knitting, sewing and reading by Mrs. Murray and her Bible-reader. I sit down among them and make friends with the well-behaved pretty little creatures, by "turning off" a garter, and hemming a few inches of a ruffle. The school began with two children and not one dollar. More than twenty are now in regular attendance, and daily means have come with daily strength for daily needs. And even in bigoted Hebron the Master has given this pair of trusting laborers great favor in the eyes of the people.

MOSQUE OF OMAR.

CHAPTER XXXVII.

THE THRESHING FLOOR OF ORNAN.

WE are so fortunate as to have in our first visit to Mount Moriah the guidance of Rev. Selah Merrill, D. D., LL. D., whose long consulate in Jerusalem and ability as archæologist and historian have qualified him beyond any other living man for the office of cicerone to the Holy City and the environs thereof. Our little party is preceded by Mohammed, the "kervasse" of the American consul, a personage so much more magnificent than his nominal master as to deserve, or demand, especial mention. From top to toe he is official. The embroidered vest; the loose sleeves, stiff with gold lace; the jauntily-ferocious tilt of the red fez capping his six feet of altitude, the wand of office in his right hand (something that suggests in equal measure the "hunting-crop" of an English squire and the "caduceus" of Mercury), go to make up the imposing presence stalking down David Street, then turning into side-ways lined with stuffy bazaars.

He takes, and keeps, the middle of the narrow thoroughfare leading into the heart of the city from the open space before the Grand Hotel. At ten o'clock on this fine winter morning all the world is abroad. The heavy rains have washed the steep street almost clean. We see more distinctly than during previous walks that it is paved with square stones, and, instead of being graded in the ordinary way, drops to the lower level every ten feet or so, in a marble step six inches deep. Of course, no wheeled vehicle can be used upon it, but donkeys, laden with human and inanimate freight, amble up and down the miniature precipices; camels lower their clumsy hulks, one foot at a time, and climb as if each step were an outrage to the inner brute. Barefoot boys belabor the donkeys' unyielding flanks; turbaned men tug and drive the larger craft of the desert.

Mohammed, the Magnificent, goes neither around nor over anything. Come what may, and go what can, he holds the right of way in the centre of the street. Under the temperate rule of the representative of a Republic, he forbears to strike man or beast. Everything makes way for the party of pale-faces, and nobody eyes us curiously or stops to stare at us.

Without the enclosure of what were of old the Temple grounds, and which is now the wall defending the precincts of the Mosque of Omar, we pause while Mohammed, running briskly forward, disappears into the barracks and presently emerges with two uniformed soldiers. Prior to the visit of the Prince of Wales and his tutor-guardian, Dean Stanley, to the Holy Land, no Christian was suffered to pass the limits of the Temple area. To-day, Jew and Gentile may visit even the interior of the Mosque, but under the surveillance of a Turkish soldier. The brace of warriors detailed to attend us clank close at our heels throughout the three hours we spend upon holy ground, and we are cautioned not to speak of Turkey by name, or to comment even in English upon the peculiar stringency of existing governmental laws and orders.

MOHAMMED THE MAGNIFICENT.

All this matters little to visitors whose thoughts are surcharged with the associations aroused by the fact that our feet actually stand within the gates of the Sacred Place. The space enclosed by the walls of the Mosque is nearly identical with that occupied by the ancient Temple grounds. It is, for the most part, paved with slabs of white stone.

"This was the Court of the Gentiles," says our guide and instructor, and when we have crossed it and gone up a step or two—"And this the Court of the Women."

Olive and acacia and karob trees grow luxuriantly in patches of unpaved soil. Mohammed breaks off and offers us green sprays to take away as souvenirs of place and hour.

Such goodly trees may have been in David's mind when he cried out in prophetic transport: "I am like a green olive-tree in the House of God!"

MOSAICS IN FAÇADE OF MOSQUE OF OMAR.

We loiter a little in the rear of the small company of Gentile strangers, to reproduce in imagination the throngs that filled the area while the Royal Preacher bowed himself in prayer with hands spread forth to heaven upon the brazen scaffold he had reared in the sight of all the congregation of Israel. "And said, O Lord God of Israel! there is no God like Thee in the heaven, nor in the earth!"

MOSQUE OF OMAR.

The mighty plateau, built up, filled in and leveled by the Wise King, overlooks valleys and hills in all directions. Solomon's Porch, glorious in white-and-gold, formerly crowned the hill on the eastern side.

"And Jesus walked in the Temple, in Solomon's Porch."

Here the lame man clung in a rapture of joy and gratitude to Peter and John until a crowd collected about the three, "greatly wondering."

The tower on the northwest corner of the enclosed area occupies the site of the tower of Antonia, from which the Roman chief captain ran down with centurions and soldiers to rescue Paul from the infuriated Jews. This, as the readers of Josephus will recall, was the last citadel held by the Jews in the final siege of Jerusalem. The Mosque of Omar, although inferior in dimensions to the Temple, is exceedingly beautiful without and within. I well recollect with what avidity I used to read travelers' stories of the mysterious interior to which none but the Moslem faith could penetrate, and how more than one curious pilgrim lost his life in the attempt to explore it in disguise. The gorgeous environment of tiles, mosaics, wrought marbles and stained glass detains us but a few minutes in our

(338) THE TOWER OF ANTONIA, JERUSALEM.

hurried passage to what would be called in music and poetry the "motif" of the superb structure. The hoary brow of Mount Moriah breaks through the tesselated pavement directly under the noble dome. A richly-wrought railing surrounds it. I thrust a reverent hand through an interstice and let it lie upon the rough surface, deaf to the voluble prattle of the trio of sacristans who insist upon pointing out the clumsy imitation of the imprint of a man's hand graven in the granite. It is a big, ungainly hand-print, and, according to Moslem tradition,

JUDGMENT-SEAT OF SOLOMON.

was left there by the angel Gabriel. To clear this story out of the way, let me say that Mohammed is fabled to have ascended to heaven from this rock, which started with him, and was held back by the angel.

"Let the fellows tell it to you," advises Dr. Merrill. "I always do this. It is the best way of ridding oneself of them. They cannot go on telling the same story for all time, even in the hope of 'baksheesh.'"

The tale repeated with an infinity of gesticulation and gibberish, the doctor kindly engages the chatterers in conversation in their own tongue, and we come back to the bare rock over which twelve hundred years agone, Caliph-Ab-el-Melek

MOSQUE OF AKSA IN GROUNDS OF MOSQUE OF OMAR.

built a Mosque. Inwardly we are grateful for the legend that has led to the jealous preservation, without coating of marble or inlaying of fine gold, of the naked top of the hallowed mountain. Beyond reasonable doubt, it was upon this rock that Abraham "built an altar, and laid the wood in order, and bound Isaac, his son, and laid him upon the altar upon the wood."

No student of sacred history and archæology disputes the assertion that the awful apparition of "the angel of the Lord standing between the earth and heaven,

BEAUTIFUL GATE OF THE TEMPLE.

having a drawn sword in his hand stretched out over Jerusalem," seemed to David's lifted eyes to hover above this, the threshing-floor of Ornan, the Jebusite.

"So David bought the threshing-floor and the oxen for fifty shekels of silver. And David built there an altar unto the Lord and offered burnt-offerings and peace-offerings."

The great altar of Solomon's Temple succeeded that erected by his father. A Christian church was reared here by the Crusaders, and for nearly two hundred years, European kings, before assuming their crowns, laid them, in solemn dedication, upon this rock. It is fifty-six feet long by forty-two wide. Dr. Merrill shows us a round hole about the size of a man's body in the heart of the huge

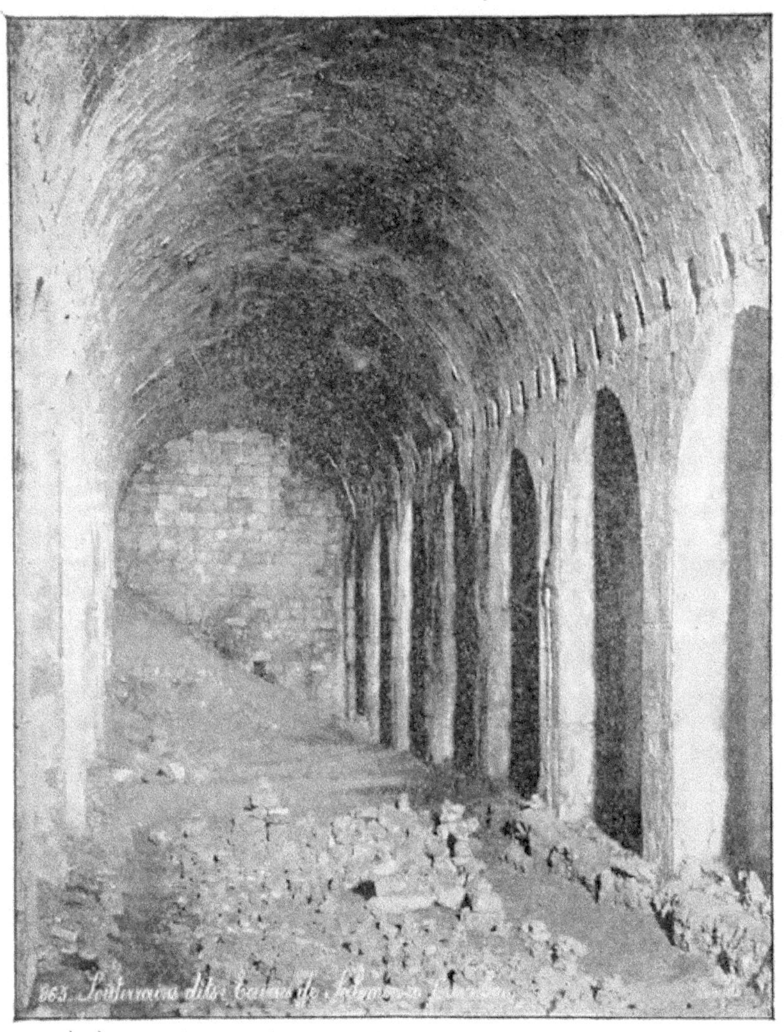

(342) (SO-CALLED) STABLES OF SOLOMON IN TEMPLE AREA.

stone, believed by competent judges to be the aperture by which the blood of victims slain in sacrifice flowed into escape pipes. A cave or crypt beneath the upper stone is floored with marble, and a circular space, corresponding with the upper opening, gives back a hollow reverberation when struck. Could this be lifted, the vast "leader" flooded daily with the typical purple tide—

> The blood of beasts
> On Jewish altars slain,

could, probably, be followed to the valley a hundred feet below. One intrepid explorer, many years ago, by the help of huge bribes, obtained permission from the guardians of the place to lift the marble cover and lower himself by ropes into the very bowels of the mountain. At the last moment, superstitious fears overcame avarice, and the permission was revoked. The Mohammedans say that the hole is the mouth of hell and that evil spirits would arise in a great cloud should the sealing marble be removed.

Directed by Dr. Merrill, we make out the location, beyond the central rock and railing, of the Holy of Holies. The space once filled by ark and cherubim is small. No devout Jew will enter the mosque, lest he should inadvertently tread upon the hallowed spot which it was not lawful for any man to visit save the high priest, and he but once a year.

In leaving the grounds, one of the party leaps up lightly to snatch for me a bit of hyssop growing in a cleft of the stones, quoting of the King-philosopher:

"And he spake of trees, from the cedar-tree that is in Lebanon even unto the hyssop that springeth out of the wall."

The traditional site of Pilate's Judgment-Hall is hard by the Temple enclosure. The howls of our Lord's enemies must have echoed through the midnight silence of the courts to which the tribes had that day repaired with "joys unknown" to celebrate the Passover-feast.

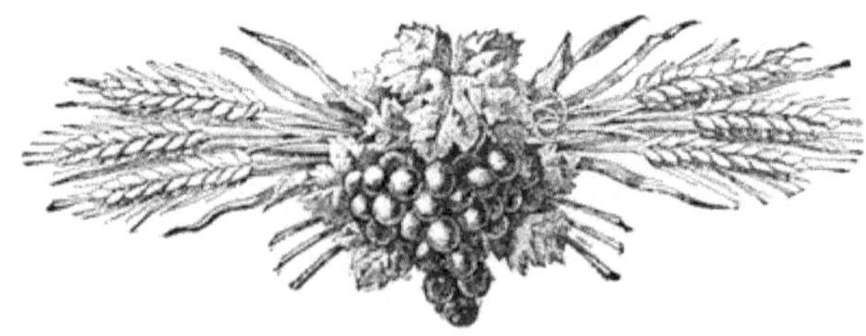

INTERIOR OF BEAUTIFUL GATE OF THE TEMPLE

CHAPTER XXXVIII.

"THE GREEN HILL FAR AWAY."

IT is Sunday morning and without communicating our intention to anyone except our dragoman we have left the hotel directly after breakfast, accompanied by him have passed out of the city through the Jaffa gate and, skirting the walls as far as the Damascus Gate, diverged there from the highway to climb this gentle eminence. The top and one side are studded with many flat and a few upright tomb-stones. For uncounted generations the Moslems have buried their dead here, and thus protected it from the encroachment of

"GOLGOTHA."

buildings for the use of the living. Across the road rises the north wall of Jerusalem, founded upon and built into the solid rock. The ancient capital of Israel was "a city that had foundations." In the native granite upbearing the massive masonry is a rude door, closed to-day and locked, leading into Solomon's Quarries.

We explored them, yesterday, marveling to see "what manner of stones" were left there by workmen who hewed out and squared the material for the temple "so that there was neither hammer nor axe, nor any tool of iron heard in the house while it was in building." Big blocks, still as white as chalk, because never exposed to air and light, bear the mark of chisel and wedge. When the quarry was first opened many years ago, bones of men and also of animals, and fragments of pottery were found in out-of-the-way corners. These last were probably

"THE GREEN HILL, FAR AWAY."

left there by workmen, while the human remains may have been those of fugitives from the law or from persecution forced to hide in forgotten dens.

Diagonally opposite the quarry yawns a natural cave, known familiarly as the "Grotto of Jeremiah," from a tradition that the prophet lived here while writing his Lamentations. I name it now because it, and another wide mouth in the upright rock, form the eyes in the "skull," which is considered by some to assist in the identification of this hill with Golgotha. By the help of imagination anybody can make out the eyes, the line of the nose, and the mouth. Some think it marvelous and attach great importance to the fantastic likeness. Others find in the round top of the hill warrant for calling it "the place of a skull." Thoughtful archæologists set aside these features of the spot, whether natural or artificial, as unnecessary in proving that here was enacted the most momentous scene in the history of the world and of the universe. Since our arrival in Jerusalem we have studied carefully the evidence of what is now the received hypothesis of the most learned and devout Bible scholars of the age, with regard to the place of the Crucifixion. As briefly as is consistent with comprehensiveness, I will rehearse some of the reasons we have for believing this "green hill" to be the true Calvary.

I have already cited several arguments against the once popular belief that the Church of the Holy Sepulchre covers the place where the Cross was set up,

"A GREAT HEBREW CEMETERY LAY ABOUT THE BASE."

as well as "that where the Lord lay" during the three days separating His Death from His Resurrection. It is needless to repeat them here further than to say that the Church of the Holy Sepulchre is not, and has never been, outside of the gate of Jerusalem, and, that Our Lord was led forth to the "place of a skull" and "suffered without the gate" is directly affirmed by the writers of the Bible. Scripture students will not require to be reminded of the important significance of this circumstance in connection with Christ our Sacrifice.

In looking, then, for a hill resembling a skull in shape, without the gate, in a public place where many passing by would witness the Crucifixion, bearing in mind also that the place of public execution among the Romans remained the same from one age to another (as witness the scene of Paul's death at Tre Fontane, near Rome, well-known as the ancient Tyburn of that city), and seeking to find records of other executions as having occurred in the same vicinity; reading that the new sepulchre of Joseph of Arimathea was near the place where He was crucified, hence, in consecrated ground already used as a Jewish burying ground —it would be strange, indeed, had intelligent explorers failed to see how faithfully all these requisites are met by the hill on which we now sit.

"THE SOLITARY TOMB."

Behind us, as we face the city and look directly into the Temple enclosure, lies what was known four hundred years after Christ as the "Place of Stoning" and as the scene of Stephen's death. It bears the name still in Jewish traditions, and there, according to Jewish chroniclers, criminals were hung up by the hands after death until the sun went down. A church in memory of St. Stephen was raised upon the "Place of Stoning" by St. Eudocia in the fifth century, the ruins of which have lately been excavated down to the exquisite mosaic pavement. In turning to the account given of the death of Stephen, we note that the marginal reference against the words, "they cast him out of the city," is Hebrew xiii. 12:

"Wherefore, Jesus, also, that He might sanctify the people with His own blood, suffered without the gate." The proximity of the two places is, to say the least, a singular coincidence.

A great Hebrew cemetery lay all about the base and upon the slope of the line of hills of which Calvary is one, for centuries before Our Lord's birth. Of this I shall speak more at length by and by. If Joseph of Armithea had wished to be gathered to his fathers at his death, he would naturally have made hereabouts his new tomb.

At "Herod's Gate" began, in the time of Our Lord, the great military road leading from Jerusalem to Caesarea Philippi, along which Paul was hurried at night by his escort of four hundred and seventy Roman soldiers, he riding in their midst upon the beast provided for him to bring him safe to Felix, the Governor. This road, recently uncovered, winds directly about the foot of this hill, and forks of the great highway lead off to Damascus and other large towns. At the Great Feast of the Passover it would have been thronged with passers-by, and lent itself fully to the custom of the Romans of making their places of execution as public as possible by establishing them near the busiest thoroughfares.

I cannot resist the temptation to quote at this point from the eloquent comment of Dr. Cunningham Geikie upon his able summing up of proofs as to the identity of the "New Calvary" with the old.

"Here, then, on this bare, rounded knoll . . . the Saviour of the world appears to have passed away with that great cry which has been held to betoken cardiac rupture—for it would seem that He literally died of a broken heart. Before Him lay outspread the guilty city which had clamored for His blood; beyond it, the pale slope of Olivet from which He was shortly to ascend in triumph to the right hand of the Majesty on High; and in the distance, but clear and seemingly near, the pinkish-yellow mountains of Moab, lighting up, it may be, the fading eyes of the Innocent One with the remembrance that His death would one day bring back lost manhood to the Kingdom of God."

I read it aloud, and then, more slowly and tenderly, the story of the Crucifixion given by each of the four Evangelists.

The day is very still; the place is deserted but for ourselves, seated upon a flat tomb-stone, and David, who stands within hearing of the familiar, always thrilling words. About our feet fragile crocuses pierce the earth, clustering closely upon the southern side of the grave-stones. Hyssop grows freely in the pale-green turf, and wild lilies shoot rank leaves above the lower herbage. Masses of clouds, white and gray, roll slowly apart, glints of sunshine stream fitfully upon the low cliffs at our right, where is another and much smaller cave than the Grotto of Jeremiah—a sepulchre hewn in the rock, and overlooking a garden. A very ancient well attests the age of the enclosed space, and olive and fig trees grow in the garden.

We let ourselves down to the lower level, and are in front of what Gordon Cummings, Selah Merrill, Lew Wallace and many other godly and learned men believe to be the long-hidden Tomb of Christ.

We have visited the vaults excavated, with the ruined church built by the pious Eudocia, and seen with strange emotion the grooves in which moved the rolling stones used in Our Lord's time for closing the mouth of the principal tomb containing several sarcophagi or niches in which bodies were laid. When uncov-

"ODDLY RIVEN FROM TOP TO BOTTOM."

ered, these were full of bones, great heaps of which are still to be seen in holes beneath the niches. A road divides this cemetery from Calvary, and the explorers have not been allowed to dig beneath the highway for other graves that, no doubt, reach all the way to the solitary tomb in the side of Calvary. A board door has been fitted into the opening once filled by the rolling stone, and a square window cut out in the same side of the rock allows anyone to look in. The door is kept locked, and the key, during Dr. Merrill's time of office, was kept at the American Consulate. A single empty sarcophagus fills the far end of the small chamber. There is no inscription, and, as I gratefully record, no frippery of altar, candles,

GROTTO OF JEREMIAH IN THE SIDE OF CALVARY.

lace and artificial flowers detracts from the solemn simplicity of the shrine. Upon the brink of the natural cliff is a great rock, oddly riven from top to bottom by the shock of an earthquake, each severed projection corresponding with a depression or gap upon the other side of the fissure. David points out these peculiarities diffidently:

"Of course nobody can say positively that it is so, and we Protestants are, maybe, over careful in such matters, but I can't help thinking whenever I look at that, of what St. Matthew tells us—that "the earth did quake, and the rocks rent."

I find the place and read on:

"And the graves were opened, and many bodies of the saints which slept arose, and came out of the graves after His resurrection, and went into the Holy City and appeared unto many."

The saints that slept in the mighty cemetery lying along these hills from Olivet to Golgotha—the vanguard of the multitude which no man can number who shall stand beside their victorious Lord at the last day upon the earth.

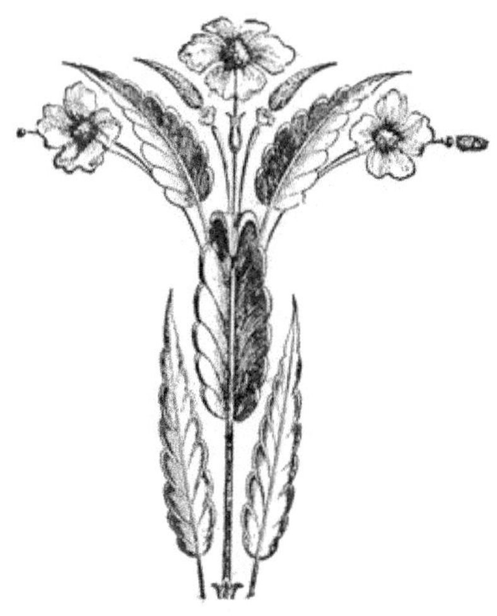

CHAPTER XXXIX.

ON THE ROAD TO BETHLEHEM.

NO other stage of our eventful journeyings has seemed so like a beautiful dream, from which we keep thinking that we must presently awake to disappointment, than this glorious December morning on which we find ourselves actually setting out for Bethlehem.

Ten months ago, when the plans for this tour were first discussed by us, we had but one definite idea as to the dates and methods of our travels:

"If we go," we said, resolutely, "we will spend Christmas in Bethlehem."

Our itinerary, as then sketched, has undergone many changes, some deliberate, others forced upon us by unforeseen exigencies. This one design has stood by us, a fixed pivot, upon which all modifications have revolved.

It is the twenty-fourth of December, and a cloudless day. Going, as is my custom, every clear morning, to the loggia on the roof of the Grand Hotel, for a view of the city and encompassing hills, I see the sky of a clear amber behind the far-off mountains of Moab; the silver-gray olives on the descent of the Mount of Olives shining softly in the rising sun, and upon the western horizon the round crown of the hill where Rizpah the daughter of Aiah watched the bodies of her dead sons "from the beginning of harvest until water dropped upon them out of heaven, and suffered not the birds of the air to rest on them by day, nor the beasts of the field by night."

"And it was told David" (here, in his royal capital of Jerusalem), "what Rizpah had done."

It is not perhaps unnatural that the vision suggested by the sight of the lonely hill-top should be joined in my mind with that of another mother, David's descendant, Mary of Bethlehem, who was to watch upon the brow of another Mount of Expiation, her soul pierced through by as sharp a sword as that which rankled all those anguished days and nights in the fierce heart of her who had borne unto her kingly lover Armoni and Mephibosheth. The joys, the pains, the sorrows and the compensations of motherhood have been the same from Eve's day until now.

There is no question as to the safety of the carriage-road to Bethlehem. A macadamized causeway runs all the way from Jerusalem to Hebron, past the birthplace of David and "David's Greater Son." Our carriage is one of many that take it to-day. The blue of the sky is like that of Switzerland; there are no mists in the valleys between the russet red and gray-green hills. Our horses are good, and the rush of the air in our faces and lungs is an ecstasy. At every hundred yards we are thrilled by the mention of names as well known and as dear as those of our own kindred and the places trodden by our childish feet.

"The Valley of Rephaim!" says our guide, indicating fertile fields bounding the track and stretching out far to right and left. We stop the carriage to "look it up."

The border-line between Judah and Benjamin was drawn hereabouts. It "went up by the Valley of the Son of Hinnom"—("There is the valley of Hin-

THE STONE OF ELIJAH.

nom!" interposes the dragoman)—"Unto the south side of the Jebusite; the same is Jerusalem. And the border went up to the top of the mountain that lieth before the valley of Hinnom westward, which is at the end of the valley" (or Rephaim) "northward."

We turn back to Genesis:

"And in the fourteenth year came Chedorlaomer and the kings that were with him and smote the Rephaim," giants and powerful, 1921 years before the birth of Christ.

"If the Captain will look at Second Samuel, fifth chapter and eighteenth verse," says David, modestly.

"The Philistines also came and spread themselves in the valley of Rephaim."

Repulsed by David, leaving their idols behind in the retreat to be destroyed by David and his men, they "came up again, and spread themselves in the Valley of Rephaim."

The figure of the rolling hosts, like spreading waters over the fertile reaches, is apt as we survey the twice-chosen battle-ground.

The exquisite episode of the "sound of a going in the tops of the mulberry-

THE HOUSE OF BENJAMIN.

trees" follows, after which David smote the Philistines from Geba until thou come to Gazer."

In turning the next leaf of what we note down as "a peripatetic Bible lesson, with illustrations from actual scenes,"—we alight and mount a stone at the road-side for an unobstructed view of the landscape we have overpast. Just here Abraham must have caught sight of Mount Moriah on the third day of his journey from his quiet home in the shadow of the grove (or tree) he had planted in

Beer-Sheba and where he had sojourned for many days before receiving the awful command to sacrifice his son—"this one only son, Isaac, whom thou lovest, upon one of the mountains which I will tell thee of."

"Then, on the third day Abraham lifted up his eyes, and saw the place afar off. And Abraham said unto his young men—"Abide ye here with the ass; and I and the lad will go yonder and worship, and come again to you."

"*Here*"—where we read the story, our backs turned upon the Mount of Promise, on our way to worship upon the Mount of Fulfillment, on the birthplace

"WELL OF THE THREE KINGS."

of the Divine and only Son whose day Abraham saw in the long, long vista of the ages and was glad!

Our guide has a tradition—frankly admitted to be only a tradition—to relate of another, and a flat stone on the other side of the road. Elijah, also coming from Beersheba, where he had left his servant, "went a day's journey into the wilderness and sat down under a juniper tree; and he requested for himself that he might die; and said, 'It is enough! now, O Lord, take away my life; for I am not better than my fathers.'"

After the miraculous refreshment of "a cake baken on the coals," and water, he "laid him down again," and slept so long and hard that he left a shallow impress of his body in the rock, says the legend. The face of the flat stone is slightly indented, but with the utmost aid of willing imaginations we cannot trace the outline of a human form. A thrifty olive-tree shades the prophet's hard bed, and near by is a large substantial building—the monastery of St. Elias.

Another monastery is pointed out on our right when we have bowled along the smooth road for ten or fifteen minutes longer. "The house of Benjamin"

TOMB OF RACHEL.

means little until the information is added that here it is said Benjamin was born on the way from Bethel, when "there was but a little way to come to Ephrath."

We are out of the carriage again, open Bible in hand, and walk up the shaded alley leading from the gate to the convent. A thrifty olive orchard hides the house from the highway, and we do not care to go up to the door of what we can see is a modern building. The sunlight flickers between the leaves over the page upon which we read how "it came to pass, as her soul was in departing (for shedied) that she called his name 'Ben-oni,' but his father called him

'Benjamin.'" The grove is sweet and peaceful; budding lilies are springing luxuriantly in the walks. We find ourselves wondering at what season the poor mother laid her down, unexpectedly, upon her bed of pain and death, and if olives and lilies grew here then.

On the other side of the way is another traditional landmark, the Well of the Three Kings, where, as the story runs, the three wise men from the East stopped to water their horses when Herod had sent them to Bethlehem to search diligently for the young child, and again beheld the star which they had seen in the East

"HEAPS PRESSING DOWN THE DEAD."

going before them, "till it came and stood over where the young child was. When they saw the star they rejoiced with exceeding great joy."

The legend may be, like a thousand others, the work of monkish fancy, but heard here, with the white roofs of Bethlehem in sight, it is exceedingly beautiful and impressive. The well, or spring, is surrounded by stonework said to belong to the Roman period.

Our next, and longest, pause in the wonderful series of object-lessons is at the Tomb of Rachel, one of the best authenticated localities in Palestine.

"And Rachel died, and was buried in the way to Ephrath, which is Bethlehem.

And Jacob set up a pillar upon her grave; that is the pillar of Rachel's grave unto this day."

"When thou art departed from me to-day, then shalt thou find two men by Rachel's sepulchre in the border of Benjamin at Zelzah," Samuel told Saul between six and seven hundred years after Rachel's burial. The inimitably pathetic prophecy of Rachel weeping for her children, quoted by Matthew as fulfilled in the massacre of the children in Bethlehem and the coast thereof, refers to the tomb of Benjamin's mother within sight of Bethlehem-Ephratah.

IN THE FIELD OF BOAZ.

A chamber about twenty feet square and as many high is surmounted by a low dome. Behind this is a sort of quadrangular courtyard with a flat roof. The entire building is of stone, put roughly together with cement. There is no attempt at decoration without, nor, as we are assured, within. This we take upon hearsay, as visitors are not allowed to enter the inner chamber. To-day, the building is closed in every part. We stroll quite around it, finding it blank, mean and desolate. The Jews are the custodians of the tomb, but the graves

TOMB OF RACHEL.

scattered all around it are of Bedouins. A few are enclosed by rude masonry, but the great majority are irregular heaps of stones, collected at random, and piled with no attempt at order. An oblong ring of the large stones marks the confines of each grave; the rest are huddled within this, making a sort of cairn.

The sight is forbidding enough before we learn that the heaps pressing down the dead are needed to protect the bodies from hyenas. Recollecting how often we have heard the horrid laughter of the unclean beasts through our tent-walls, we shudder at the weird scene conjured up by fancy by this new scrap of knowledge. Of the bare and lonely sepulchre, dim under the stars, or defined by the moonlight, the midnight silence broken by the growls of the baffled brutes as they tear at the heavy stones, or shriek with delight over the rifling of a shallow grave into which the tenant was thrust with careless hands, and left unguarded except by the gravelly sand we crunch under our feet. It is a dreary resting-place for the patriarch's petted wife. The only vegetation near it is a woolly-leaved plant growing under the wall. I pluck a spray and press it in my note-book. Beneath I pencil the touching digression from the main line of Jacob's narrative to Rachel's elder son, forty years after he had laid his darling down for the long sleep on the roadside. After speaking of Joseph's sons, Ephraim and Manasseh, as entitled to a place among their father's brethren, rather than among his issue, the old man begins to wander backward in his talk, and never completes what he had begun to say at the first:

"And, as for me, when I came from Padan, Rachel died by me in the land of Canaan in the way, when yet there was but a little way to come unto Ephrath; and I buried her there in the way of Ephrath; the same is Bethlehem."

Did something in Joseph's eyes bring up the face of his beautiful mother to the dying husband who had never ceased to be her lover? Or had the father a vague intention of commending to the son the charge of her grave, and the pillar Jacob had set upon it?

CHAPTER XL.

CHRISTMAS IN BETHLEHEM.

BETHLEHEM is approached by a succession of hills, terraced to afford foothold for olive, fig and mulberry-trees and vineyards. In David's youth it was a prosperous region, and the successors of Boaz and Jesse maintain a tolerable reputation as good farmers. By contrast with the shiftless husbandry between Jerusalem and Jericho it is like a garden in order and fertility.

The appearance of the town is imposing when we behold it from the plains below. The group of churches, about which the white stone houses are clustered, have some architectural pretensions, and rise finely from the summit of the ridge sloping away from them on two sides. When surrounded by walls the site must have been picturesque, and as a fortification, strong. At a distance the houses look higher than they really are, being built upon the sides of the ridge. After

ON THE ROOF.

reaching the town we perceive that it is not an exception to other Syrian "settlements," of whatever size, in point of commodiousness and cleanliness. The steep streets are the only sewers, and are so narrow that two carriages cannot pass in many of them. They are also so crooked that, as our vehicle rattles around a corner, the driver gives a warning yell to pedestrians lest they should be surprised and hurt. He dashes through one and whirls into another in great style, cracking his whip and shrieking out his cautions to the occupants of doorways and the middle of the thoroughfare, while the foul mud flies in every direction. The peo-

A COMPANY OF GYPSIES.

ple take him quietly, with true Oriental indifference, flattening themselves against the walls and retiring into open doors and arches until the danger is over. In one very strait passage we meet two camels, the nose of the second tied by a rope to the saddle of the first. We graze the mire-encrusted side of one and nearly upset the other, without lessening our speed, until we draw up with a mighty clatter and jerk at the door of the "Hotel Bethleëm."

The great square between the hotel and the Church of the Nativity is filled with people, most of them in holiday attire, and after a hasty luncheon brought with us, but eaten in a big, bare room of the caravanserai, we go up to the roof for a better view of the striking scene. A parapet, three feet high, runs all around our look-out. There are a few chairs for infirm or distinguished guests, and plenty of standing-room for a party little less varied in character than the greater assembly below. Five nationalities are represented among the spectators upon our house-top, the hotel being beneath us. As far as we can see them, every roof is filled with people, the white veils and gay jackets of the women giving a festive air to the multitude. The aspect of the open square—as I reflect with a thrill of emotion—probably does not differ materially from what might have been seen

"ONE END OF THE MARKET-PLACE."

there almost nineteen hundred years ago on Christmas-Eve. "All the world" was to be taxed, and Bethlehem Ephrata was little among the thousands of Judah; too strait to accommodate the return of those of the house and lineage of natives of the town, who must be enrolled there.

There will be no room in this, or any other inn of Bethlehem to-night for one-fourth of the visitors drawn hither by the prospective celebration out-of-doors and in the church. In the shadow of the barracks opposite a company of gypsies have pitched their tents and are cooking their dinner in the open street, an operation gravely superintended by an ass in harness. A bearded shepherd and his wife have lambs in their arms for sale. A string of kneeling camels occupies one end of the market-place, their gowned and turbaned owners chatting, or, more likely, chaffering near by. An old man and a girl who pulls one end of her izzar over her mouth, as a man in European cut-away coat and Derby hat turns to look at her in passing, sit upon the ground with half a dozen hoppled hens at their feet, awaiting purchasers. While the clamor and bustling activity of an American gala-day are absent, the subdued hum of many voices arises to our ears like the swarming of a thousand bees.

Mary had come up from Nazareth with her husband and would be weary of nerve and limb, diffident in the presence of the motley crowd, anxious to get into shelter, and disappointed, with the heart-sickening chagrin of a modest peasant girl, when she found that there was not a vacant nook in the khan where she could lie down and rest. We make it all so real to ourselves that we resent the complacent well-being of the loungers below, the greater comfort of the watchers upon the secluded house-top. We visited the oldest khan in Jerusalem last week, and it helped us to understand into what sort of quarters she was compelled to retire at last. The little light came through openings in the groined roof; the stone floor was littered with bundles of straw and hay along the four walls, which had fastened to them raised mangers but a foot above the ground. We have likewise seen several stables in caves, and learned that these natural grottoes are, in some parts of Syria, preferred to buildings above ground.

"UPON THE ROOF OF A LOW WING OF THE BARRACKS."

Our minds are preoccupied with the Jewish maiden while the panorama of shifting figures and colors passes before the field of vision The crowd grows more dense; from the barracks issues a company of Turkish soldiery in brilliant uniform and marches down the middle of the square, throwing the masses of people right and left until a broad track is left clear from the end of the street to the entrance of the church. Upon the roof of a low wing of the barracks is collected a knot of men sumptuously appareled, to whom an officer evidently reports now and then.

"The Patriarch of Syria and Palestine is too infirm to come in person, and he has sent a Bishop who will be here presently. All these people are waiting to see his arrival. The Pacha could not come, and has sent his dragoman. It is a case

"THE GORGEOUS PAGEANT."

of substitution, all the way through," somebody says in English, and sneeringly.

The Bishop is lunching somewhere, we are told vaguely, and, apparently, is in no hurry to quit the table. Until he comes, we may not enter the church of the grotto of the Nativity. But for the novelty of the scene and the liberty to sit silent if we choose, and think out our own thoughts, we should be intensely weary before a blare of trumpets sounds down the steep street and is answered by the bugler in front of the barracks. A mounted officer gallops into the square, striking with the flat of his sword at trespassing groups. Two foot-soldiers, armed with whips of hippopotamus-hide take, each, one side of the way, and literally *lash* the crowd back into place, respecting neither sex nor age. We can hear the hiss and thud of the heavy whip, and are amazed that no outcry, much less resistance, follows the blows. The populace are used to the summary methods of their masters, and take it all as a matter of course. Into the area thus rid of the lower orders of humanity, rides a small body of horsemen; several carriages come next; the military band bursts into triumphal music; the distinguished party leaves the roof and a majestic figure descends from a coach upon a carpet spread for his feet. At the sound of the trumpet, a procession of priests and choir-boys emerge from

"EVERY HOUSE-TOP IS FILLED."

the church and, advancing slowly, singing as they go, have reached the carpeted space as the bishop steps from the carriage.

Now ensues a ceremony inexplicable to us, and, perhaps on that account, fantastic to ludicrousness. The black robe with scarlet hood worn by the prelate is dexterously whisked off, and a purple robe with a train several yards in length substituted. Gloves of the same color are drawn upon his hands, which are then folded upon his breast and kept there, fingers extended stiffly like those of an automaton. Thus dressed, the bishop moves toward the church, attended by scores of ecclesiastics and acolytes, all wearing black gowns, and above these the garments we style, for the lack of the technical term, white dressing-sacks trimmed with lace that looks cheap, and is dear. Censers swing clouds of perfume into the air, the sustained chant of men's voices mingles with the chimes from the belfry; women drop upon their knees as the gorgeous pageant passes them. The bishop's pace is inconceivably deliberate; his steps must not be over an inch long.

"Like our little bride!" says a laughing voice in my ear.

My Syrian interpreter, to whom I am indebted for the story of "Fudda's Wedding," stands close behind me, her black eyes full of enjoyment of my surprised interest in the performance. Four choir-boys bear the purple train; the white-robed attendants sway in the slow march like banks of lilies in a gentle breeze; the air is vibrant with bells and voices. It is all imposing and beautiful, and the pulses keep time to the throbbing of chant and chime.

My interpreter leans near.

"I wonder"—the careful articulation of a foreign phrase making the sentence more significant—"*I wonder what Jesus thinks of all this!*"

CHAPTER XLI.

CHRISTMAS IN BETHLEHEM (CONTINUED).

THE Church of the Nativity was built by Constantine (305–337 A. D.) and restored in the eleventh century by the Crusaders. The pillars upholding the roof are of reddish stone, and scratched all over with the names and crests of these valiant knights:

> "Their swords are rust,
> Their bones are dust,
> Their souls are with the saints, we trust."

Eight centuries have, in passing, rubbed into illegibility what each man wrote of himself and his deeds in this high and holy place. More distinct, and easy to be read by a Greek scholar, is the inscription upon the octagonal baptismal font:

"*A memorial before God and for the peace and forgiveness of sinners (whose names the Lord knows).*"

One wall bears in (now) shattered mosaics rude pictures of the churches of Sardis and of Antioch. There were many other churches represented here and some mosaics date back to the twelfth century. Beyond the immense columns of dusky-red, the shell of the church with huge rafters richly embrowned by time, and the mutilated mosaics, little remains of the temple reared by Constantine and piously remodeled by the Crusaders. Modern ecclesiastical art has crowded the interior with costly decorations, most of which are in good taste, although some are tawdry.

The service of Christmas Eve begins at half past ten at night.

After a very poor dinner served in the Hotel Bethlëem, and eaten in the good company of half-a-dozen Americans, an Englishwoman, a German officer and an Italian gentleman, we disperse to our several bedrooms and try to sleep until the hour of service. With myself the attempt is utterly vain. It is not only that the stone floors and walls of the small chamber create a chill that pierces to bone and marrow, and that an unclosable window opens upon the staircase, down which boots are clattering incessantly. My head is hot and my brain awhirl. In all the day I have had but one still hour in which to pull myself together and appreciate in heart and soul where I really am, and why I have come. At sunset I got away from the crowd of people and press of distractions, and found my way to the roof alone.

CHURCH OF THE NATIVITY, SEEN FROM BELOW.

The great square was no longer a restless mass of human life, but was dotted with groups of quiet figures; the area about the gypsy camp was faintly illumined by the wavering flame of thorns crackling under the supper-pot; here and there a duller glow revealed a brazier over which the owner warmed his hands or boiled his coffee. Beyond the massed houses with their twinkling windows, the outer darkness was rolling up from the valleys. The hills lay hushed and low against the cold yellow of the horizon, and the stars were kindling in the zenith. The air was breezeless, but frosty, and I paced the wide area until the sky was full of stars,

THE CHURCH OF THE NATIVITY.

keeping watch over Bethlehem and the surrounding land, as "shepherds watched their flocks by night,

All seated on the ground,"

upon the gentle slope over there, still bearing the name of the Field of the Shepherds.

It is of that lonely rapt hour, never-to-be-forgotten, and not to be described by pen, that my mind is full to the exclusion of thoughts of slumber. In all the meditations of the day, the Galilean peasant-girl has moved, a presence that m[illegible] felt, almost seen. Her *Magnificat* is read again, and for the third tim[illegible]ay, before we set out for the church.

It is thronged to suffocation. David has had a man on the spot, holding chairs for us ever since sunset. He lies, half-asleep, across them when we appear to claim them. The aisle upon our right is filled with kneeling women; the long white linen veils worn alike by by maid and wife, give a wondrously picturesque appearance to the throng. Men stand, shoulder to shoulder, in the aisle upon our left; the Europeans bare-headed, Turkish residents of towns wearing the red tarboosh, the shaven skulls of fellaheen bound with turbans, generally white. A liberal sprinkling of soldiers, on police duty, enlivens the more sober garments. The music is fine from choir and organ; at least forty gorgeous ecclesiastics, in gowns stiff and glittering with gold embroidery, are within the chancel-pale. The bishop

AREA ABOUT THE GYPSY CAMP.

sits in the tall patriarchal chair at the extreme left of the high altar, as we face it, the brilliant array of churchly millinery and altar-furniture is seen mistily through clouds of incense.

It is half-past eleven o'clock, and the organ still rolls; priest has relieved priest in the nasal sing-song reading of the "First and Second Gospels" in Latin, anthem has succeeded chant, the bishop has been undressed and re-dressed twice, as to slippers, robe and headgear. The lofty, steepled hat, gleaming with gold and precious stones, has been placed over his white hairs and taken off again so many times that we have stopped counting. The chairs are hard; the air is stifling; except for two student-like men near me, who speak English and are probably perverts to the Roman Catholic Church, nobody in the audience, that I çan see, makes any pretence of following the service. These men read silently all the while from

INTERIOR OF CHURCH OF THE NATIVITY.

Latin Testaments, and when I ask Mrs. Jamal in an undertone the meaning of some manœuvre of the ecclesiastical forces, one turns to offer me a book taken from his pocket.

"Perhaps you would like to read the Gospels as they go on?"

Declining with thanks, I note that his accent is American, that he wears a strait-breasted, long-skirted coat and a plain, high collar fastened at the back of his neck. His companion is as evidently English.

Ten minutes of twelve! Many men have sat down upon the marble floor, crossing their legs under them and gone to sleep; among the white-veiled ranks

GREAT SQUARE AT NOON.

of kneeling women I see some who have had the wisdom, or good luck, to get close to the great pillars and now rest tired heads against the cool stone, eyes fast shut—perhaps in devotion. A significant pause in the peal of the organ and responses from the two lines of choir-boys causes a stir in the crowd. Indifference and drowsiness are exchanged for alert interest. Raised heads are turned toward the high altar, above which I observe for the first time a curtain hung before what may be a niche or a painting, perhaps two feet long, and half as high. After a low prelude, the organ glides into a lullaby, sweet and tender and tuneful, that has in it the very swing of the cradle, and the brooding love in mother-heart and the mother's voice. It is still sighing through the now strangely-quiet church when the curtain hiding the niche above the altar is drawn back by an unseen cord or hand, and we see—a cradle, with a doll lying in it!

To those about me what is to me an anti-climax, belittling the august preliminaries, and inconceivably unworthy of the occasion that has collected the vast audience—is apparently impressive and affecting. Men stand on tiptoe to gaze; many bow their heads, and some kneel as if overcome by emotion; women are convulsed by sobs, or smile lovingly through tears. Not even the glorious *Gloria in Excelsis* that rolls a triumphant volume of sound up to the embrowned rafters that have been jarred by the Angels' Song for over a thousand years, brings back to me one symptom of the feeling I have striven all day long to encourage, in spite of Bishop's toilettes and a host of other (to me) senseless mummeries. In the abrupt revulsion from thoughts of Mary the handmaid of the Lord, and of the Birth, which was a world's redemption, I find myself sick in body as in heart, and were I not wedged in the crowd to an extent that would make extrication a work of time and disturbance, I should beg to be allowed to leave the church.

As the drawn curtain and the pealing *Gloria* proclaim the Event of the evening, the packed aisles are agitated by an arrival. Two amazingly-appareled kervasses bearing long, gilded staves, force a passage for a trim, natty little gentleman, military in bearing and in the white "imperial" tuft upon his chin. Keen-eyed and suave, he is conducted to a chair immediately in front of us.

"The French Consul!" whispers Mrs. Jamal.

The Roman Catholic Church in Palestine is under the immediate protection of France, and the homage paid to the handsome little man may savor of gratitude. Sitting erect and portentously attentive, a gorgeous kervasse upon each side, he is the most distinguished personage present, even before two priests part themselves from the crowd of ecclesiastics and approach him while the choir-boys chant melodiously. One of the priests bears what looks like a gold plate about six inches in diameter, and bending respectfully offers it to the lips of the consul, who arises at his approach and kisses the gold disk, whatever it may be. Somebody explains that it is "the Seal," but what Seal or why thus presented and saluted, I am unable to learn. In equal ignorance as to the meaning of the marching and countermarching; the kneelings and uprisings, and especially the robings and the unrobings of the majestic Chief of the hour, I sit for two-and-a-half mortal hours longer, a miserable fixture in my uncomfortable chair. The Third and Fourth Gospels are intoned in Latin by a third and a fourth glittering priest; all read and chant through their noses, and it is within bounds to assert that not fifty out of the hundreds assembled in this, the oldest church in the Holy Land, comprehend one syllable of what is uttered.

And still the white-veiled women continue to kneel in the aisle, their faces turned altarward, and the motley-hued garments of the men are pressed close together in the body of the church.

"They will not stir until the Baby is carried around," says my interpreter, when I wonder that the crowd does not thin.

At half-past two, "the Baby" is borne in solemn procession of all the priests, amid chanting and organ thunder, and the adoration of the now wide-awake and excited people, down one aisle and across the end of the church, then up another

FIELD OF THE SHEPHERDS.

aisle and back to its resting-place near the high-altar. It is a large wax doll, gowned in lawn and lace, not a particularly pretty, but a very artificial baby. It goes without saying that we do not prostrate ourselves as it passes, but there are not many other exceptions to the rule of what may not be worship in very deed, but is so much like it that the uninitiated are excusable for drawing no distinction between the two.

In the crush of the retiring spectators at the door, I am hustled against Mrs. Sharpe, and she seizes my hand in both of hers, speechless with emotion.

I remark in a voice that must sound jaded, that I had seen Dr. Sharpe across the church, but without her, and supposed her absent.

"I have passed the whole evening upon my knees in the Grotto!" she sighs, hysterically. "I was never so blissfully happy before in my life."

Arrived at the hotel, which is full to overflowing, I go at once to my room, and to bed, cold, disgusted, disappointed, and, I doubt not, disagreeable to my patient attendant who does her best to mitigate physical malaise by a cup of hot tea, and by a hot-water bag tucked between the chilly sheets.

At intervals during the hour that elapses between bidding her "Good-night" and falling asleep, I hear the measured fall of footsteps upon the roof overhead, sounds that recall, with soothing efficacy, my twilight promenade and reverie.

In the morning I learn that Alcides has walked there, and alone, under the silent stars until half-past-three.

"To restore the balance you know," he remarks, concisely.

I do know!

CHAPTER XLII.

STILL IN BETHLEHEM.

THE Grotto of the Nativity is reached by a stone staircase of thirteen steps leading down from the church. It is a natural cave, and the church, having been built directly over it, shuts out the light of day.

Before visiting Bethlehem, we have carefully looked into the evidence tending to substantiate the genuineness of the claims of the grotto to the high honor bestowed upon it by the Christian world. A few sentences on this head may not be amiss here.

One of the earliest authorities who has left on record any information as to the birthplace of Our Lord was Justin Martyr, whose generation overlapped that of the Apostle John. St. John lived to be nearly one hundred years old, and Justin Martyr, less than one hundred and fifty years after the birth of Christ, says explicitly that his Lord was born in a cave very near to the village of Bethlehem. This Christian Father of the early Church was a native of the ancient Shechem (now Nablous), and very possibly may have himself talked with John. It is nearly certain that he must have known people who recollected hearing in childhood the wonderful story of the Messiah's miracles, death and resurrection from their parents who had been eye-witnesses of these things. Every student of Roman history as associated with the early life of the Church is familiar with Hadrian's daring impiety in planting a grove and building an altar to Adonis over the grotto in which the Jewish peasant, whose followers were turning the world upside down, was said to have been born of a virgin. This was done between 117–138 A.D. Fifty or sixty years after Hadrian's death, Origen, another of the early fathers, writes that even the pagans acknowledged the cave to be the place where Christ was born. In the fourth century, the saintly Jerome took up his abode in a neighboring cave that he might study and write, live and die, as near as might be to the spot of his Lord's Incarnation. A church had already been built upon the exact site once desecrated by the idolatrous rites commanded by the Roman Hadrian.

When the Crusaders occupied Bethlehem, they found Christians there treasuring with pious zeal the tales handed down to them of the honors lavished upon their town, the city of David, and the birthplace of David's lineal descendant. There are still families in Bethlehem who pride themselves upon the intermarriages of ancestresses with the Crusaders, and upon the superior physical and mental strain that survives to this day in the offspring of these unions.

In all ages, and among all sects of Christians, there has no been doubt as to the truth of the statement that Our Lord was born in the place where we now stand. The exact spot is said to be the arched recess directly before us. The first thought with the devout visitor is invariably a regret that the natural walls of rough rock appear only in two or three parts of the cave, and, especially, that the shrine of shrines is overlaid with colored marbles and further adorned with gold and silver—a display that absolutely vulgarizes it. A silver star let into the marble floor of the semicircular niche is kept bright by the tears and kisses of pilgrims. We do not kiss it, but we kneel to lay a reverent hand upon the symbol of the Light of the World, the Star of Bethlehem.

RUSSIAN PILGRIMS.

"*Here Jesus Christ was born of the Virgin Mary*," is the translation of the Latin inscription.

We are sorry, on some accounts, to have visited it for the first time at a season when throngs of devotees make quiet meditation impracticable. We have to move aside almost immediately to make way for those who are waiting for their turn to prostrate themselves and press their lips to the star. A crowd of Russian pilgrims have walked up from Jerusalem this morning. Most of them are peasant women, past middle age, and it goes to my heart to see them kneel, mutely, one after another, here and touch the holy spot with forehead and mouth, kissing it passionately, over and over, sometimes sobbing quietly, yet never uttering a word.

GROTTO OF THE NATIVITY.

Light from a semi-circle of silver lamps suspended by silver chains shows the star, the vari-colored marbles, and above, the embroidered lambrequin hung from a marble shelf furnished with its usual appurtenances of a Roman Catholic altar. This is guarded by a gilded grating from careless or sacrilegious touch. We are thankful that the space beneath has no such defence.

We bestow but a glance upon the so-called Manger, disfigured by ecclesiastical millinery, and fenced about with gilded wires behind which swing more silver lamps. Overhead, angels hold a scroll inscribed in Latin with the angel's song: "*Peace on Earth, Good-will to Men.*" We care even less to linger at the altar erected where, as the Church has it, the Three Wise Men adored the Holy Child. The sacristan shows us, also, the place of Joseph's warning dream and the crypt in which the slaughtered Bethlehem babies were buried.

BACK-STREET IN BETHLEHEM.

We do not pretend to believe in the authenticity of any, or all of these legendary localities, but we listen interestedly to something Mrs. Jamal has to say apropos to this crypt. It is said that there are always more boys than girls born in Bethlehem, and that they are stronger and handsomer than the boys of other Palestine towns.

"Because the dear Lord will make up to Bethlehem women through all the ages for what they suffered and lost when the Innocents were killed."

It is a beautiful thought, and we cannot deny that we have, before hearing the

MANGER IN CHURCH OF NATIVITY.

tradition, remarked upon the number of fine-looking urchins playing in the streets and attending the services of the church with their mothers.

The cave tenanted by St. Jerome is close at hand and in it his tomb is shown with that of Paula, the disciple to whom he refers in his works.

As we come thoughtfully out of the upper church, David directs our attention to the venders of crosses, beads, painted candles and other mementoes of Bethlehem, who have stalls or tables in the vestibule, and even hawk their wares just inside of the doors. There are actually money changers just outside, ready to take up and exchange foreign coins for Turkish money.

DAUGHTER AND DAUGHTER-IN-LAW.

"It was such as these that Christ our Lord drove out of the Temple, declaring that they had made it a den of thieves," he utters in grave disapprobation.

We go in quest of David's well under the conduct of our worthy dragoman, who takes us to three, all having equal claims upon our credulity. We refrain from mentioning the unsatisfactory search during a visit we pay to an old inhabitant of Bethlehem and a chief among his people, who will have us drink a tumbler of water "from the well of Bethlehem which is by the gate." There are several mouths to the one spring of living water upon the ridge on which the city is built, and there are no signs of a gate there now, but the old inhabitant may be in the right. He is an imperious, yet jolly and kindly patriarch, and has a history. In earlier and troublous times he exercised a sort of sheikship in the territory around Bethlehem, having such influence with his lawless neighbors that he was frequently chosen to escort travelers through doubtful passes and over hills that had more than doubtful reputation. As David tells me this in English, the host catches his meaning and points proudly to his gun, sword and spear hung against the wall.

THE SAD-EYED BRIDEGROOM.

Christmas is a universal holiday, and the retired potentate is in high feather, receiving our party and other visitors. He sits cross-legged in a corner, a snowy turban setting off his swarthy face, which is strong and intelligent, and issues orders to each member of the household staff. His wife, his married daughter and a daughter-in-law are at his beck and call, and a relative who chances to drop in, is bidden to pound the coffee roasted by the ex-sheik himself, after one woman has brought it, another the shovel in which it is to be browned, and a third has blown up the fire in the charcoal brazier before him. With palpable enjoyment of the little bustle of preparation, he turns the hot grains into an olive-wood mortar which is, he tells us, an heir-loom in his family, and scolds the guest good-naturedly, but dictatorially, for not beating the "grinding-tune" in proper time against the seasoned sides of the vessel. His daughter has a Madonna-face; the daughter-in-law a beautiful. Yet, as Mrs. Jamal tells me aside, the handsome son had set his fancy upon another woman, and even dared to reason with the parent who ordered him to espouse the village beauty.

"Marry her, or leave my house!" rejoined the father. "You will obey me, or be no longer my son."

We agree privately that the family despot is not a bad type of Jesse in dignity and authority, and the impression deepens with each minute of our stay in this better-class household. The son is now very happy in his father's choice, we are told, and that his sad eyes are the expression of bodily illness,—a sort of low fever he cannot shake off. He smiles pleasantly, and even proudly, as the bride of three months' standing, to please us, retires to her room and puts on her wedding-dress, returning blushingly for our inspection. The gown is of red and green silk, in alternate "gores" of each color, put together with a sort of herring-bone pattern done in red and green. Over this is worn an embroidered jacket, parting over a vest wrought in many stitches and colors. Upon her head is a stiff, helmet-like cap of green-and-red woolen stuff, made upon a pasteboard frame, and bordered above the forehead with overlapping rows of gold coins. Strings of silver "pieces" depend from the sides and join below her chin in one large gold coin. These are an important part of her dowry, and after marriage she seldom lays off the heavy head-dress, even sleeping in it.

I exclaim at this:

"I should think it would make your head ache!"

"It did, at first. Now, my head aches if I do not wear it."

The weight of the cumbrous thing is materially increased by a linen veil, richly wrought at the ends with silk thread in an intricate pattern. It is three yards long and a yard wide. The texture and needle-work of these veils vary with the means and station of the wearer. Girls wear veils of linen or cotton, but not the stiff construction beneath it.

The host got a good price in exchange for his Madonna-faced daughter when he bestowed her in marriage, but had to pay a larger for his son's wife, who is reckoned as yet comelier by kindred and neighbors.

VILLAGE OF ABOU-GOCH, ON THE ROAD FROM JERUSALEM TO JAFFA.

CHAPTER XLIII

JAFFA.

BOTH the carriage-road and railway from Jerusalem to Jaffa run through Samson's country, the stamping-ground of the long-maned, strong-backed, soft-hearted Nazarite. A village of new, neat farm-houses, one of Rothschild's colonies, is in the immediate vicinity of Zorah, the home of "a certain man of the family of the Danites whose name was Manoah," and the birthplace of his son. A large rock, lately uncovered, is supposed, from indications that it was used as an altar, to be that upon which the Danite offered the kid unto the Lord:

"And it came to pass when the flame went up toward heaven from off the altar, that the angel of the Lord ascended in the flame of the altar, and Manoah and his wife looked on."

Samson's range in his fightings with the Philistines, and his courtings of divers women, lay between the hills on either side of the road. Zorah, Ekron, Gath, Kirjathjearim, Gaza, the rock Etam, Lehi—are names that suggest the years upon years of warring with the original lords of the land who were a chronic pest to the conquerors. The House of Dagon at Ashdod has left no trace upon the landscape of to-day; Ashdod itself is represented by a few poor mud huts roofed with straw and turf.

Gezer, made a City of Refuge by Joshua, and afterward taken by Pharaoh of Egypt, rebuilt and given by him to his daughter, one of Solomon's wives, crowns a hill on the right of the road. Far up the heights on the left is the cave of Makkedah (also lately discovered), in which Joshua cooped the five kings until the battle was won, when "he smote them, and slew them, and hanged them on five trees," afterwards casting their bodies into the cave and laying great stones upon the cave's mouth, "which remain until this very day."

From the brook (marginal reading "valley," now a wide, dry water-bed, paved with loose white pebbles), crossed more than once by our track, David took the five smooth stones for his sling, and somewhere in this same valley was the great fight between Saul and the Philistines who "stood on a mountain on the one side, while Israel stood on a mountain on the other side, and there was a valley between them." It is impossible to designate the exact spot where the stripling confronted Goliath of Gath, but we select a locality to please ourselves and fight the battle over to our satisfaction.

HOUSE OF SIMON, THE TANNER, IN JAFFA.

"Lydda was nigh unto Joppa," when Peter, tarrying in the smaller town with resident saints, healed Eneas who had been bed-ridden for eight years,—in the sublimely simple formula—"Eneas! Jesus Christ maketh thee whole. Arise, and make thy bed!"

He was still a sojourner in Lydda when summoned to Joppa by the news of Dorcas's death. The village is brought so near to the seaport by rail that a touch of New World enterprise would make it a suburb easily accessible for merchants who care for purer air than is to be had in the narrow streets of modern Jaffa. The Crusaders set their stamp upon Lydda in the form of a noble church built above

OLIVE GROVE NEAR JAFFA.

the alleged burial-place of St. George, the patron saint of England, and coupled in everybody's mind with the dragon. The remains of this structure, dating from the twelfth century, are amiably shared by the Greek Church and the Moslems, part of them being used as a mosque. Besides this building there is nothing of interest in the squalid village.

The approach to Jaffa is heralded by groves of orange and olive trees. Bleak hills, where the only signs of human habitation are villages of low stone houses, the materials being taken in many cases from the ruins of citadels and walled towns, make way for sand-dunes drifted from the seashore by winds that make the harbor of Jaffa a word of terror all over the world. Fertile patches fight with these—sometimes victoriously, sometimes with such indifferent success that the fringes of grass and starved-looking grain edge oases of vines and fig-lands, and end abruptly in yellow sand. Then, the orange groves thicken into orchards, encircled by hedges of prickly-pear with big, fleshy leaves and ugly stalks as crooked as snakes and as large as a man's arm, and we plunge suddenly into a labyrinth of alleys, lined with dirty stone houses, standing flush with the pavement—the usual, and now the old story in Syrian towns. One of these conducts us to the Jerusalem Hotel, a rambling building of various ages, over the modest doorway which we read, "*Go through, go through the gates!*" Gardens lie behind

CAMELS LADEN WITH JAFFA ORANGES.

it, and by the way of these we go to see the house of Simon the Tanner, where Peter lodged. While neither of us believes that the unpretending building, evidently not more than five hundred years old, and which may be less, is that once occupied by the apostle's landlord, it is quite possible that the site may be the same. Tanning has been the business of the locality for perhaps twenty centuries, and the always restless surf of Jaffa booms against the sea-wall bounding the court-yard. We pay to enter this, the fee being taken by a man in the garb of a Moslem priest, the care-taker of the shabby mosque of which Simon's house is an appanage. Another man, similarly attired, lowers a skin bucket into the old well in the yard and gives us to drink of the water. It is exceedingly brackish and

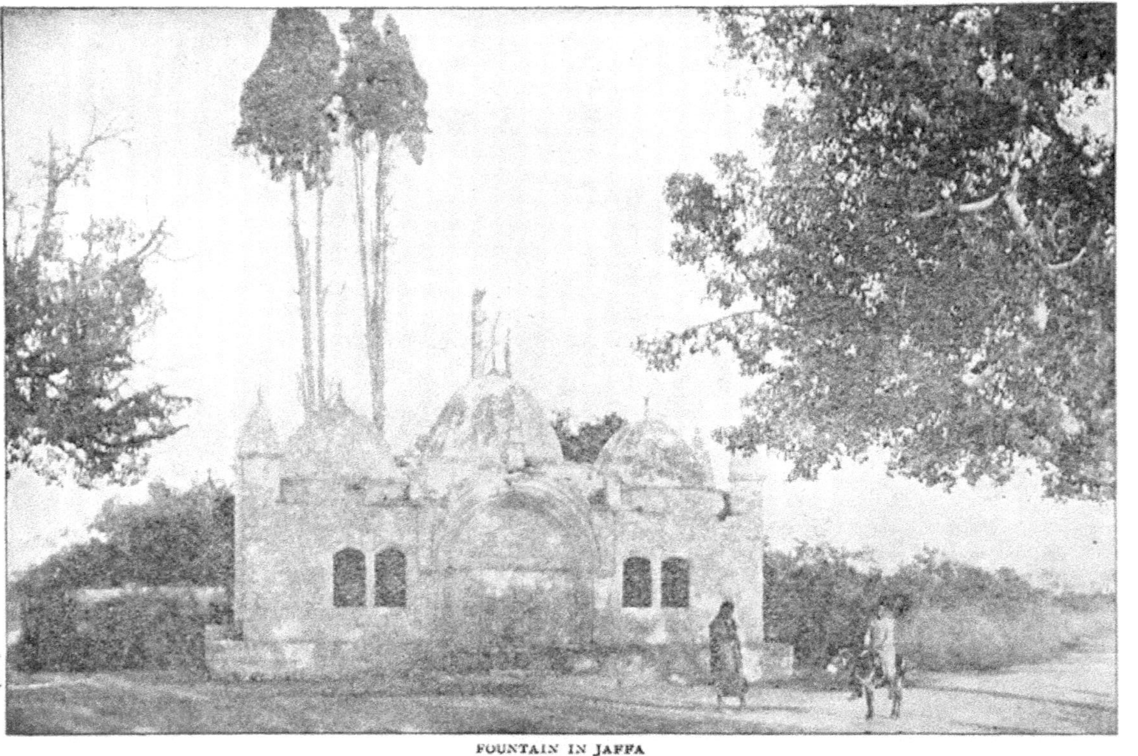

FOUNTAIN IN JAFFA

salt as might be expected. We pluck a few leaves from the fig-tree overhanging it and ascend the short staircase built upon the side of the house, to the flat roof. It is surrounded, like the pretty nest on the house-top we visited in Jerusalem, with hollow tiles, arranged horizontally in geometrical designs and set in cement, for the convenience of Moslem women who would peep at the world. A parapet of some sort was about the house-top chosen by Peter as an oratory, "a battlement for thy roof" being one of the laws of Moses. The fisherman-disciple certainly never fell into a trance upon these cemented stones, yet the scene spread out for him

STREET IN OLD JAFFA.

that noon-tide, before the miraculous vision locked his senses against external objects, was like in general features that upon which our eyes now fasten eagerly. The great and wide sea—the "utmost sea" seen by Moses' dying eyes—the Mediterranean, blue, seductive and treacherous of mood and meaning, rolled and thundered then against a jetty built by a Roman Emperor. The ruins of this now make dangerous the always difficult business of landing and embarkation. The hills, the distant line of mountains, the ineffable blue of the smiling heavens, were beautiful then as they are now. But the town had in Peter's time a wall and

JAFFA (JOPPA) FROM THE HARBOR.

a Roman garrison, and commodious inlets where had been moored Hiram's floats laden with Lebanon cedar for the Temple in building at Jerusalem.

"And Majarkon and Rakkon with the border before (over against) Japho," (Joppa) were deeded to the Danites in Joshua's grants of Palestine lands. From Joppa, Jonah sailed for Tarshish (probably in Spain) instead of to Nineveh. The territory and town still belonged at that date to the Phœnicians, from whom the Hebrews had never been able to take them. Joppa has been a bone of contention among "Turks, infidels and Jews," and enough blood has been shed in defence and conquest of this rough key to the Mediterranean to dye the blue waters crimson. In using the figure we look toward the east at a group of sand-dunes pale-yellow in the sunlight, that should be forever red. There Napoleon—the great Napoleon—had three thousand prisoners shot in platoons, to avoid the trouble and expense of transportation to Egypt.

On our way to the hotel we meet a funeral procession; turn aside to let it pass and then follow the line of march through the very muddiest streets we have ever seen, even in Palestine. Not one person in the company of mourners has shoes, or even sandals on his feet; the men wear abiehs and turbans, the women gowns of dark-blue cotton, and white (by courtesy) izzars. They are not too much afflicted to hold these up, we perceive, when the puddles reaching clear across the street are unusually deep. The bier is borne upon the shoulders of four men. Our guide makes us notice that it is an unpainted open box, so shallow that the breast, shoulders and profile of the corpse are visible from the ground.

"He was poor, and the friends cannot afford a new coffin. This has been used many times already. Rich people have new and handsome coffins, which are sometimes buried with them. You hear that they sing happy songs," continues the speaker, a Jaffa man, and an acquaintance of David. "That means that he was a good man. He made a pilgrimage to Mecca but lately, so short a time ago that he has, the Moslems think, surely gone to the bosom of the Prophet. He has not had time for much sin since. So nobody cries, and the hymns are glad."

Maybe so, to an ear cultivated into an appreciation of Syrian melody. To us, the happy hymns are as dolorous as dirges, and like all other Moslem music, tuneless, but for a rhythmic beat of time such as a drum-stick brings from a drum in the intervals of march or quick-step. The chanting of Southern negroes in antebellum days, of long sentences all upon one or, at best, two notes, was something akin to the nasal sing-song to which the motley crew splash through pools in the lanes and the stiffer mire of the market-place, where not a man or woman turns out of his way to make room for them—up to the door of the mosque, where we leave them.

Our solitary call in Jaffa is upon Miss Arnott, the principal of one of the best schools in the Holy Land, now over thirty years in successful operation. Since leaving Beir..., we have not sat in a room that reminded us of home so forcibly as the pleasant parlor where we are entertained by the true gentlewoman and philanthropist to whom is due the good accomplished by this great light in a dark place. David Jamal has two daughters here, one engaged in teaching where she was herself taught, the other as a pupil. All the native assistants are graduates of the school, and the gentle dignity of the principal is reflected in the deportment of her subordinates. Dismissing David, and sending Alcides with him, I have a long and, to me, deeply interesting talk with Miss Arnott. She left home and country in her youth to devote her whole life to the task of lifting her fellow-women out of the depths wherein they have lain for ages of despotic rule and unspeakable degradation. She has laid hold of the only hope for her sex in Syria, and toiled hopefully for the girls who are to be the wives and mothers of the next generation. After all these decades of patient labor and manifold discouragements, she sees her school a recognized power in the land and sits, a queen-mother, among her grateful children-in-heart.

MARKET-PLACE IN JAFFA.

CHAPTER XLIV.

DRAGOMANS AND HOTELS.

NEXT to health and the resolution to make the best of everything, a good dragoman is the most important factor in the sum of Palestine travel. As the country becomes more accessible to the visitor from foreign lands, the supply of guides and interpreters increases. I should like to add that preparation for the profession is more thorough than when there was not one-tenth as many who followed it.

The first and indispensable qualification for the position of dragoman to American tourists is, of course, a knowledge of the English language. The number of schools established throughout Syria has greatly facilitated this object, and made it an easy matter, moreover, to get a smattering of history and archæology that impresses the superficial observer with the apparent intelligence of the native displaying it. The respectful attention with which he is heard raises his opinion of himself, already inflated by the consciousness of superiority to his uneducated fellow country-people. He can speak English, and usually French, as well, and they only Arabic; he wears better clothes than they and knows something of European ways of living and behaving. In his capacity of guide and interpreter, he is continually consulted by his party, and a tolerable degree of familiarity with routes and the history of the places visited exalts him into a personage of importance. But for him the tourist could not make a purchase, or exchange his gold for Turkish coins, or so much as ask for a drink of water. Being, as a rule, a man of less than ordinary breeding to begin with, a son of the people, his head is turned by his new position, and he assumes—not unreasonably—that, like the storied "little ostrich," he "knows it all."

Such specimens of the genus dragoman, flashily-dressed in the uniform of the guild, swagger in the corridors and ante-rooms of hotels, and exchange loud greetings in the market-place, and haunt places of interest on, or off escort duty, rattling off statistic and tradition, and directing the movements of "parties" with assurance becoming appreciation of the little brief authority inseparable from the office. The duties of a dragoman, as the better-behaved and better-informed of the class comprehend, are to conduct the tourist in safety and comfort over the proposed line of travel; to guard him from imposition; to cater to his tastes, not only as to food, but in the style of entertainment offered him, and to give him full and trustworthy abstracts of the history and legends associated with each place visited. In all this, he is, nevertheless, to hold himself subject to the wishes and fancies of those he has in charge. However important in his estimation it may be that the party shall visit this ruin, or see that form of amusement peculiar to the country, it is his part, having once laid the case before them, to abide cheerfully by their decision, whether to follow his respectful advice or their own caprice. He is paid to wait upon their pleasure, not to play the dictator or master.

Our own experience with our incomparable dragoman prepared us to criticise with surprise and indignation what we characterized as the "freshness" of others

"OUR INCOMPARABLE DRAGOMAN."

we chanced to meet. I well remember the glow of resentment with which I turned upon a young fellow to whom, while David was busy elsewhere with my son, was deputed the duty of showing me the way back to the hotel.

"Have you other sons?" he amazed me by inquiring presently.

"No," I answered, stiffly.

"Won't you take me to America with you? I would be another son to you."

"You will please call one of those donkey-boys over there," was my only reply, "I will ride the rest of the way."

In narrating the incident afterward to a resident of Jerusalem, he explained what I took for insolence by telling of several instances of formal adoption of good-looking young dragomans by wealthy Americans, subjoining incidents that transferred my displeasure from the vain, but ignorant, native to the traveling American. It is not unusual, according to this informant, and others who have had equal opportunities of observation of the relations of tourist and guide, for travelers from our free and independent country, to insist that the dragomans should sit at the table with them while in camp, and to treat them in all other respects as equals, and even as honored guests. American girls will walk arm-in-arm with favorite dragomans in the streets of Damascus, Beirût and Jerusalem; run races with them on the plains and down the hills, divide an orange or apple with a nice-looking Syrian guide, eating half and giving him the other; eat bonbons from the same box, and—I blush to say—smoke cigarettes with them. Sometimes, when the guide is especially prepossessing and entertaining, the Daisy Miller of oriental journeyings goes so far as to carry on a lively flirtation that amuses her and utterly spoils the native of a country where the frank, innocent association of the sexes is absolutely unknown.

Before visiting the Holy Land, the traveler should obtain through trustworthy sources the address of a really competent dragoman—one conversant with every step of the proposed tour, honest and intelligent, and who will make the cause of his charge his own from the moment he enters upon his duties. It is the habit of many—as we found to our cost in Cairo and Alexandria—to make common cause with merchants who pay them commissions upon every sale arranged by the dragoman. After two months of David Jamal's jealous watchfulness over our interests, it was a shock when our guide to the Pyramids and the Sphinx protested almost angrily against our call at a shop near both, where we wished to buy souvenir spoons.

"My travelers are never allowed to do their own buying," he asserted. "There was Mr. ——, from Buffalo, New York, who bought five hundred pounds' worth of things in Cairo, and left the whole business to me. I choose them, I settle on the price, I pay the money, and he ask not one question."

Upon learning that the money spent by us in Cairo would probably not exceed five pounds, he became sulky and washed his hands of the whole transaction. The traveler is a veritable victim in the hands of such a man, powerless to protect his pocket and hardly able to defend himself against insult. He is wise who, at the first sign of arbitrariness, declares his rights and his intention to maintain them.

"How do you like your dragoman?" I inquired of a Prussian officer whom we met in Bethlehem.

I had noticed the fellow in Jerusalem and been disagreeably impressed by his smirking effrontery and the bullying tone used to a party of unsophisticated American women and children.

"Oh, he is quite tolerable if I kick him (figuratively) three times a day. Unless I do he is simply unbearable."

Before positively engaging a dragoman, one should, by letter, arrange every particular of the journey, settle everything relative to a time, expenses, etc., and have a written contract drafted, to be signed by both parties. The arrangements for the tour which we have just happily completed were made six months prior to our arrival in Syria, and I record with satisfaction that every provision of the contract was fulfilled to the letter and in spirit. Not one unpleasant word was ever spoken by one of the parties concerned in the expedition; we never saw a sullen look upon the face of an attendant, or failed to get a ready and courteous response to any request.

After the frequent tributes paid in these chapters to the worth and services of our dragoman, it would seem superfluous to recommend him in terms of unqualified approbation to any who may have in view a visit to the Holy Land. Mrs. Oliphant, the English authoress, in the preface to her book of Syrian travel, alludes to David Jamal as "the Providence of our little party." He also conducted Canon Tristram, the Prince of Wales, Dean Stanley, and other eminent personages through Palestine, and always with the same result of perfect satisfaction to the traveler. Of our obligations to him personally I have not room to speak, but I must cite him as an acknowledged authority, even among scholars, in all pertaining to the history, the ruins and antiquities of his native and beloved land. His Biblical knowledge is wonderful; his reverence for holy things modest and unaffected; his character for integrity and honor unimpeachable. His address is simply, "*David Jamal, Jerusalem, Syria.*"

It gives me pleasure to speak here of his friend and late partner, Demetrius Domian, as an excellent guide and intelligent dragoman. Were I to revisit Syria and could not get David Jamal, I should undoubtedly endeavor to secure Domian's services. His home upon the house-top, overlooking the Valley of Jehoshaphat and Mount of Olives, is the prettiest nook we found in Jerusalem.

He who expects to see in any Syrian caravanserai the elegant "conveniences" of Parisian and American hotels, is doomed to sore and certain disappointment. The nearest approach to these was found in the Grand New Hotel in Jerusalem, where the ingenious thoughtfulness of Mr. Gelat, the manager, goes far toward compensating for the lack of appliances for properly heating in cold weather a house built with express reference to long, intense summers, and for the impossibility of procuring fresh milk, butter and fruits. It is within bounds to say that no one can suffer other than minor inconveniences under his care.

The Hotel d'Orient in Beirût, while inferior to the Grand New of Jerusalem in furniture, fare and general comfort, is considered the best in that city.

The Hotel Dimitri in Damascus outranks the d'Orient of Beirût in sanitary provisions, in the variety and preparation of food, and in the appointments of table and chambers.

Everywhere men are employed as chambermaids, an awkward circumstance in sickness, and indeed at all times when the travelers are women, and unaccompanied by their own maids. After a while one gets used (comparatively) to the queer order of service, as likewise to the deprivation of many refinements she once esteemed essential to a moderate degree of bodily well-being.

In bidding farewell to the patient reader who has kept pace with me to this final "Talk," I must blend with thanks for his courtesy and forbearance, grateful acknowledgments to a few of the many whose counsels and timely assistance in the hour of need have filled up deep places, leveled hills, and led us by ways of pleasantness and paths of peace, back to home and friends.

To Mr. Louis Klopsch, of New York, whose was the inception of the project of our Oriental tour, we owe more than to any other man who forwarded the long-coveted end. His wise foresight provided for our convenience and pleasure at every stage of the journey, and each arrangement was made upon a scale of kindly generosity which is beyond praise.

To the Faculty of the Protestant College at Beirût and their families for loving hospitality to the weary stranger within their gates; to Rev. Joseph Jamal, of Jerusalem; to Dr. Sandreczky, of the same place, and, in an especial degree, to Rev. Selah Merrill, D. D., LL. D., late U. S. Consul at Jerusalem, our thanks are due now and always, for a wealth of kindness and friendly aid that leaves us for ever and hopelessly in their debt.

With us the pleasure and interest of our eventful jauntings "up and down in Palestine" so far outweighed the mishaps, that the latter are reduced in the retrospect to a minimum,—delicate shading thrown in here and there, to make brighter the lights that will never go out until Eternal Day breaks for us upon the shore beyond the river, and we know for ourselves the fadeless glories of the New Jerusalem that to mortals have never been told.

CHAPTER XLV.

MY FRIENDS, THE MISSIONARIES.

MY opposite neighbor at table upon the voyage from New York to Southampton in the autumn of 1893 was a young woman about twenty-five years of age, whom I silently decided by the closing of the second day out, to be among the most interesting of my fellow-passengers. In feature she was pleasing—even pretty—but her charm lay in a certain refinement of speech and manner, combined with quick intelligence and sensibility of expression. She was a lady in grain, and in education and conversation so far above the average of her age and sex, that when the crucial twenty-four hours of "slight unpleasantness" to both of us were happily over, I made opportunity to cultivate our acquaintanceship.

We were already good friends when, on the fourth night of our voyage—which chanced to be Sunday night—we were pacing the moonlighted deck together, and the talk took a personal, semi-confidential turn. The initiative step was my statement that I was bound for Palestine, the Promised Land of my life-long dreams, never before visited by me in body and in truth. My companion listened with flattering interest, and when I proposed jestingly that she should join me in Jerusalem, smiled brightly.

"In other circumstances, nothing would give me more pleasure, but I, too, am going to a Promised Land. My destination is Rangoon."

"Are you going alone?"

"Alone—so far as human companionship is concerned. The friends with whom I was to have sailed left America a week ago. I was detained by a short, but severe illness."

This was the preface to the story I drew from her in the intercourse that ripened fast into intimacy on ship-board.

From childhood, she had known that she was "appointed," as she phrased it, to the Master's service in foreign lands. With the natural shrinking of youth from privation and toil, she had tried to get away from the conviction in various ways, entering, at twenty, upon the study of art as an experimental lesson in the expulsive power of a new affection. At twenty-three, she was impelled to reveal to her mother the struggle going on between conscience and expediency, and how she could not escape from the persuasion that the Divine will urged her to consecrate herself to the life of a foreign missionary. The mother's reply set the seal upon her purpose:

"Were I fifteen years younger I would go with you. As it is, let me fulfill *my* part of the mission by giving you up cheerfully."

From that moment, the deep peace that entered the daughter's soul had never known a cloud. There was no sentimentality in language or ambition. She was not a dreamer, much less a fanatic. A clear-headed, resolute woman, she knew what she had undertaken, and counted the cost to the minutest detail. In putting her hand to the plough, she had grasped it, not hastily, but with staying power in the hold. In our long and earnest talks upon the subject, I appreciated for the first time, what constitutes "a call to the Mission Field." Since then I have thought and spoken of it with reverence, as something with which a stranger to such depths of spiritual conflict and such heights of spiritual enlightenment as hers, may not intermeddle.

My last glimpse of her was in the Waterloo Station, London. We had said "Good-bye" upon the train and I was seated in a carriage awaiting the disposition of our luggage, when she passed, under the escort of a clergyman who had met her upon our arrival. She caught sight of me, stepped hastily to the open door of my carriage, and the electric light showed the ineffable white peace of the smile with which she kissed her hand to me silently and made a slight, but eloquent, upward motion. Then, the crowd and the London night swallowed her up, and I saw her face no more.

The recollection of her had much to do with the resolution that moved me, a month later, to seek an interview with a party of missionaries who, I heard accidentally, were voyaging with me upon a P. & O. steamship bound to India, *via* Port Saïd. The information came to me through the lips of one of the ship's officers who was my *vis-a-vis* at table. "A jolly game of cards had been disturbed the night before by the psalm-singing of a pack of missionaries in the second cabin" he growled, "If they had sung something jolly, don't you know, the card party would not have minded it so much, although there was such a lot of them that they make a beastly racket—but hymn tunes have a way of making a fellow low in his mind—don't you know?"

I had never heard until then of missionaries as second cabin voyagers, and the impression was unequivocally disagreeable. It is not agreeable down to the present hour, although I have learned how common it is for the Board at home (moved, presumably, by the churches at home) to economize in this way, especially when the voyage is long. My readers may not sympathize with the indignation that flushed up to my forehead at the coupling of the words "missionaries" and "second cabin." It may be that the failure to fall in with my temper arises from ignorance of the conditions of a six weeks voyage, second class, in a P. & O. steamship. The first cabin passage was inconvenient to discomfort to one used to Atlantic floating palaces. The linen was dingy and musty; the food badly cooked and carelessly served; the general debility of the milk and the sustained strength of the butter were matters of popular complaint. Nothing was up to the American standard of prime quality, except prices.

I had looked over the rail separating the first from the second cabin, and marveled at the brave cheer of the less fortunate in their cramped quarters. Tell-tale odors from the kitchen and over-crowded sleeping cabin reached our olfactories; children played and wrangled and romped, like beasts in a circus wagon, all day long.

It was easy to account for the preference of men and women in all weathers for the strait bounds of deck-space allotted to them above remaining in doors.

So, the flush still burned my forehead when, as soon as breakfast was over, I betook myself to the end of the ship where was located the second cabin, and, passing through the gate, asked a ruddy young Englishman if I might have speech with my friends, the missionaries. He was one of them, he said, pleasantly, and he had the whole band about me in a few minutes. Sixteen of them—all from Great Britain. Four Wesleyans, four Baptist, four from the Church of England, and four Congregationalists. My exclamation at the equal allotment to each denomination raised a laugh, and we were no longer strangers. In breeding and education the women were the superiors of those who lounged in sea-chairs under the double awning amidships, and murmured languidly at the heat and length of the voyage. Not a man of them was as scantily equipped with brains and behavior as the officer whose game their sacred music had disturbed. Fancy-work and plain sewing went on while the men took turns in reading aloud; family prayers and bible readings were held in the cabin at stated hours, three violins and a flute made a tolerable orchestral accompaniment to the vocal music of the evening; ship's coil and shuffle-board in a measure compensated for the lack of the long promenade of the first cabin.

The cheerful contentment of the party was to me, astonishing. With one accord, they overlooked discomforts until they became glaringly obtrusive, then laughed at them. In American phrase, they were "in for a good time," and they had it. When questioned, all pitched the stories of personal experience in one and the same key. Of their own free will, and, after mature deliberation, they had entered upon a course they hoped to continue while life should last, and they rejoiced and were glad in it. Six of the sixteen were veterans in the foreign field; five were the children of missionaries, who had been educated in England, and were going out to carry on the work begun by their parents. The peace that passed worldly understanding was not the serenity of ignorance. They knew what they were undertaking.

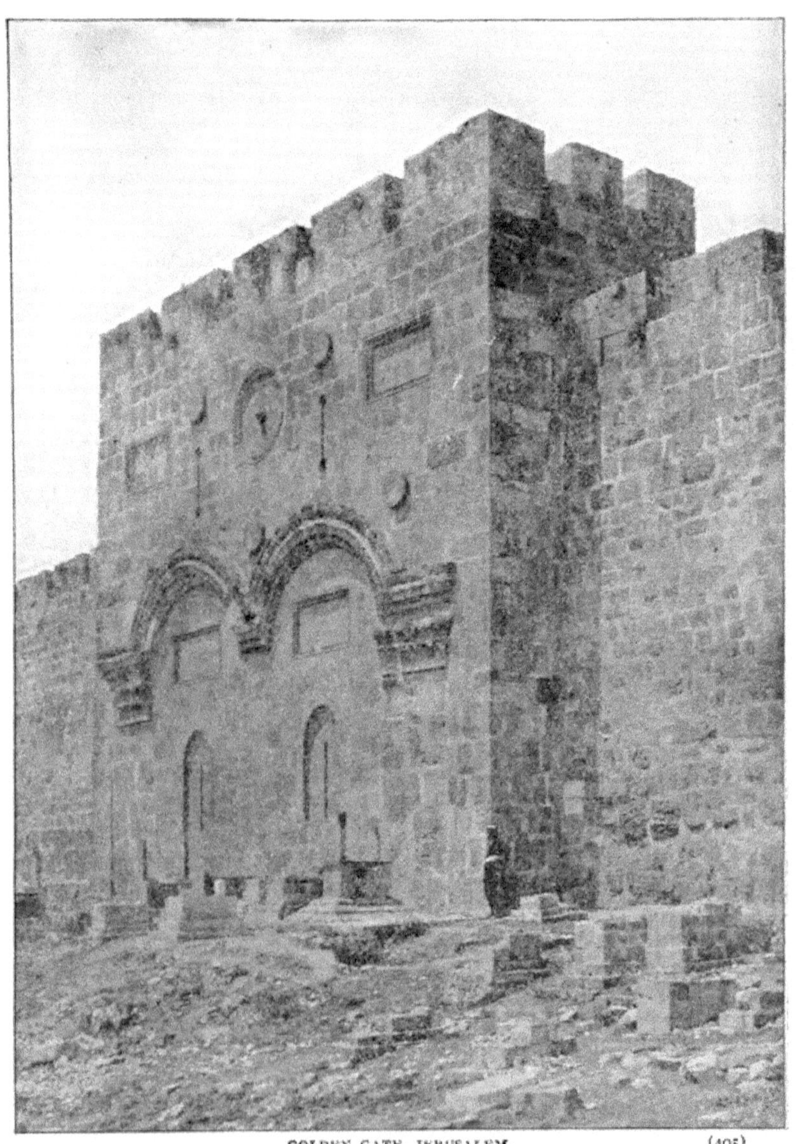

GOLDEN GATE, JERUSALEM.

A young man—a first-cabin passenger—who had heard with mingled wonder and cynicism the report of my visits to the "psalm-singers," one day asked permission to accompany me. Being a gentleman, he quickly affiliated with the missionaries, and made the most of our call. It was evening, and, after bidding them "Good night" we walked the deck for awhile, he glancing, at each turn, at the group seated in the moonlight within the cabin doors. By-and-by, he gave, without prelude, his solution of the mystery of the happiness of such people in such circumstances.

"They must love *Him*"—reverently raising his cap—"*very* much."

In six words he had furnished the key to conduct that baffles the adepts in secular policy. It is a key that adjusts itself to every combination.

Through the silence succeeding the unexpected remark, I seemed to hear, in the rush of the south wind that blew softly and the wash of the Mediterranean waves,—like the rhythm of a Gregorian chant:—

"*For I am persuaded that neither death, nor life, nor angels, nor principalities, nor powers, nor things present, nor things to come.*

Nor height, nor depth, nor any other creature shall be able to separate us from the love of GOD, *which is in Christ Jesus our Lord.*"

In Beirût, Syria, I counted my friends, the missionaries, by the score. Dr. Post, the head of the Medical Department of the Protestant College in that place—which is, to all intents and purposes, a university—was our fellow passenger from Port Saîd, *via* Jaffa, and the first hand-clasp I had after we anchored in the Beirût offing, was from Dr. Bliss, the president. For ten days and more I was in hourly association with the noble body of professors and tutors, who, with their families, make up one of the most charming social circles it was ever my privilege to enter.

During one of the calls with which Dr. Bliss honored me, he said with the air of a man who celebrates a happy anniversary:

"Thirty-seven years ago to-day I left my native land for this place and work."

"Have you never regretted it?"

"Regretted it! In looking back to-day, my regret is that I have not, in the course of nature, thirty-seven years more to devote to the same cause."

"We are sometimes spoken of as the gilt-edged mission," he continued. "But there are black edges to certain leaves of our history."

This introduced a deeply interesting abstract of the early struggles of the mission band—then a feeble folk,—against half-hearted backers at home, Moslem distrust and jealousy, the apathy of the native population for whose sake personal ease and selfish ambition were killed all the day long. I had from an eye witness the particulars of the massacre of Christians by the Druses in 1862, and of the imminent danger of the American Mission. Of Dr. Bliss's ineffectual petition to

GATE OF ST. STEPHEN, JERUSALEM.

the English Consul for *one man* and an English flag, that the missionary might plead in person with the murderers. How every native Christian man and boy in the settlements near Beirût was killed, and the women and girls were brought down from the ruins of their homes to fill to overflowing the Mission House, and be fed, nursed and clothed by the missionaries and their wives. Of an alarm of approaching peril that led to the flight by night under cover of the cactus hedges lining a lane running down to the pier, where lay a boat ready to convey the hunted American Christians to an English man-of-war. Babies were snatched from their beds and borne off by their parents, everything else of value being left for the pillagers. Of Mrs. Bliss's sigh, as she sped along in the midnight at her husband's side,—"If we could only escape to the mountains!" and his reply, "GOD is our only refuge and strength, my dear. Look at the mountains!"

The Lebanon range, that at sunset had been as the Garden of the Lord in terraced luxuriance of vine and olive and fig tree,—was lurid with the glare of burning villages.

"*Now* we have no hardships!" was said to me so often that I inferred the completeness of the missionaries' expatriation. Time and custom had, I supposed, reconciled them to the *rôle* of men without a country. The ties of land and home and kindred were weakened by the protracted tension upon patriotism and natural affection of absence and other interests.

While confessing with shame the folly and injustice of the conclusion, I believe that the comparatively small number of American Christians who have ever given this side of the mission question a thought, have fallen or slidden into the same misapprehension. My opinion was reversed by the events of the Thanksgiving Day I passed in Beirût. At eight o'clock in the evening,—an hour corresponding as nearly as could be computed with the one o'clock dinner time at home—there was a social gathering of the garrison at the house of Professor Porter. I shall never participate in such another celebration of our national festival. Addresses were made, prayer was offered for the far away native land, and we all sang as clearly as swelling hearts and aching throats would allow, "My Country, 'tis of Thee!"

I comprehended that night what patriotism, pure, simple and utterly disinterested, is like; a fire of devotion that had stood the chill of years, the creeping damps of foreign life and of interests utterly dissociated from American institutions. A patriotism absolutely divorced from politics is a phenomenon worth going half-way around the world to see.

I diverge from the main line of my theme to relate in this connection an incident of Dr. Bliss's visit to England in 1864, when the financial condition of the Beirût Mission and the distress of the parent land made an appeal to British Christians imperatively necessary:

FRONT OF THE HOLY SEPULCHRE, JERUSALEM.

At a meeting of the friends of the mission, held in a London drawing-room, Dr. Bliss announced that he had raised $10,000 toward the sum needed to put the college upon a stable foundation.

A jeering voice called out,—"In *money*? or in Yankee greenbacks?"

Without the pause of a second, the reply rang back—

"I shall not use one cent of this amount until every dollar of the ten thousand is worth a dollar in gold! Nor shall I have long to wait?"

He kept his word to the letter, and, as he had predicted, he had not long to wait.

This is the stuff of which the men are made who have set the Beirût College and Mission upon the hill commanding the harbor, the stretch of the blue Mediterranean on the left, and, across an arm of the sea, the glory of Lebanon. I can compare this station to nothing more aptly than to the colossal figure whose uplifted torch gleams nightly over New York Bay. Like her, it proclaims liberty to the captives and the opening of the prison to them that are bound by the trammels of a false faith, and the tyranny of a besotted government.

"You wonder at *our* contentment?" said one of the women missionaries to me. "I will show you a stranger thing if you will go with me a day's journey up the country."

Let me take you, who now read, with us.

Right in the heart of the hills upon the outskirts of the miserable Syrian village, is a house built of rough stones laid up in mud, and with a thatched roof. It differs from its neighbors mainly in having three rooms, where the others have but one. In it live an educated man and woman with two little children. In this mountain-parish, the American missionaries are school teachers, hospital nurses, preachers and physicians, laboring with hand and head from year to year, sometimes seeing no white visitor for months together; straitened for means to provide for the needs of the primitive household,—yet never so much as cast down, much less in despair. It is in a home like this that one enters into the fullness of the pledge—"*My peace I leave with you; my peace I give unto you. Not as the world giveth, give I unto you.*"

"Not as the world giveth!" Else they had long ago been left comfortless—in exile, poor, toiling and destitute of all solace of friendship, social intercourse and intellectual stimulus.

They were very cheerful, and very busy—this devoted pair, and inexpressibly thankful that the native women began to keep their homes cleaner, to be willing to have their girls taught to sew, and cook and read, and that a few men listened respectfully to such simple Bible stories as every child brought up in a Christian home knows by the time he is five years old.

Not long ago I met an American Congregationalist, one of whose friends had, in a Syrian tour, spent a night in this hospitable hovel.

GENERAL VIEW OF THE VASQUES OF SOLOMON.

"She thought them very good people," said the traveler's friend patronizingly, "and they seemed to have their work at heart. But she was disappointed to find them using really *lovely* china and *solid* silver forks. All wedding-presents, she said, or sent by her mother since, but such show of luxuries hurts the cause of Christ. It isn't like giving up all for Him, you know. And this is what the foreign missionary *must* do."

Of this style of renunciation of other people's goods for Christ's sake and the Gospel's, on the part of well-to-do church members, dwelling at ease in American Zions, I shall have something more to say in considering the life of the home missionary.

I pass on, now, to the last glimpse of my friends, the Syrian missionaries.

In ancient Hebron, within a quarter-mile of the Cave of Machpelah where lie buried Abraham, Isaac and Jacob, Sarah, Rebekah and Leah—we visited Mr. and Mrs. Murray, English people, and, with the exception of one other family, the only English-speaking household in the town. Mrs. Murray is blind, her husband is lame, and, when divinely directed, as they firmly believed, to this spiritual desert, the very stronghold of Moslem bigotry, they knew not one religious organization to which they could look for the means of carrying on their proposed mission. They have lived by the day a life of trust that casts into the shade any other I have ever heard of. Mrs. Murray and a Bible reader have collected a school of twenty-five or thirty little girls, whom they instruct in all sorts of handiwork, in the rudiments of letters, and in knowledge of the Bible. At the vintage season, almost the entire population of Hebron live for two months in booths in the vineyards, and the English missionaries go with them, helping the mothers to look after the babies, nursing the sick, and, altogether, making themselves one with the working people.

Mrs. Murray spoke with devout gratitude of the favor they have found in the sight of the Moslems of both sexes, although the inhabitants of the region are notoriously the fiercest in their bigotry and fanatical jealousy of any other faith to be found in Palestine.

"We have never been allowed to want for any good thing," said the blind woman, the light of a great peace upon her face "GOD has mercifully never let us doubt that this is our place in His great and wide vineyard. With this persuasion, labor in the foreign field is a blessed cross-bearing, for the Master carries the heavier end."

At the American Mission in Cairo, I had the privilege of knowing, somewhat intimately, the laborers who have made strong the foundations of a worthy enterprise. In the Bible-class of young men taught by Miss Harvey (now Mrs. Robertson), I met, besides native converts, a dozen or more young fellows in the scarlet uniform of the British soldiery, most of them Scotchmen, to whom the church service and Bible-class are like home voices, powerful in restraint and in consolation. The English occupation of Northern Egypt has made the care of this element of the motley population an important branch of evangelistic work.

Here again, was the same, and by now, the old, old story of peace that flowed like a river, and happiness in a life which, to the unlearned in such matters, appears harsh and painful, and oftimes barren of desirable results in man's impatient calculation of profit and loss.

In this cursory retrospect, I have, with intentional catholicity, dealt with various denominations of those who love our Lord and Saviour, Jesus Christ, in sincerity and in truth. Of my friends the missionaries in Jerusalem, those connected with the Church Missionary Society of London and the two gentlewomen from our own country, who at their own charges, are doing such work among the lowest class of Jews as the Murrays are carrying on among the Moslems in Hebron, I cannot even begin to speak. What I know of them personally—their toils, their faith and patience, their sublime confidence in the promises to him that overcometh—would consume, in the telling, more time than my readers have to give, or I the strength to take.

In our age, as in that in which our Lord lived and taught, the children of this world are more cunning than the children of light, but the wisest children of light are the ardent spirits that turn their backs upon the homes they love, the refinements that have become as necessary to comfort as the breath they draw, and, deaf to lures of earthly gain and honor, devote life and talent to the service of Him who established both Home and Foreign Missions in the general order that has never been repealed, and will never be outlawed, until time shall be no more.

"*Go ye into all the world, and preach the gospel to every creature—beginning at Jerusalem.*"

If this be not disinterestedness of the highest order, then I do not know what disinterestedness means. If this be not altruism of the stamp that came into being on the first Christmas day, then heroism, and self-sacrifice and the love that vaunteth not itself, doth not behave itself unseemly and never faileth, are but empty names.

P. S.—Since this book was written news has come to me over two seas of the death of one of these devoted women, Miss Robertson, of whom I have spoken in the chapter entitled, "The Box Colony." To the first impulse to regret the loss to those to whom she ministered, and to the friends who loved her, succeeds our solemn thankfulness that her unsealed eyes have looked upon Him for whose coming she watched as those who wait for their Lord.

"Does not your heart fail you, sometimes, in this daily round of duty to the miserable and unbelieving?" I asked at our last interview. "Are you never homesick for the 'Old Kentucky Home far away?'"

"Sometimes—when I am *very* tired, I am homesick, but not for Kentucky, or for America," with the sweet smile that transformed a worn face. "Then, I pray—maybe impatiently—'Lord! how long?' and 'Come quickly, Lord Jesus!' Usually, I am willing to abide His own good time."

She knows now, having entered into the joy of her Lord, why she—and the world—have been kept waiting.

The Story of Armenia

The Christian People of Ancient Eden and Their Persecution by the Moslems.

ARMENIA AND ITS PEOPLE.

SINCE the foregoing pages were written, and after Marion Harland's return from Bible lands, Asia Minor, Syria and Turkish territory in both Europe and Asia generally have been the scenes of events of a most startling character. Above all others, Armenia, the ancient Eden, and the seat of what is probably the oldest known form of Christian belief, has been visited by persecution and massacres of such appalling proportions and frightful inhumanity as to recall the early Christian sufferings under the Roman rule, when multitudes perished in a single day. Rome's enormities, however, have been rivaled if not eclipsed by the horrible outrages

THE DARDANELLES.

recently perpetrated by the Ottoman power in the plains, and on the valleys and hillsides of Armenia, where nearly one hundred thousand men, women and children of the Christian faith have been slain in cold blood—many with the most dreadful tortures, and from three to four hundred thousand others rendered destitute and utterly helpless. Furthermore, this gigantic holocaust, with all its attendant horrors of flame, rapine and violation, has continued unchecked, under the very eyes of the so-called civilized powers of Christian Europe. Whatever pangs of conscience may have assailed individuals or communities after each successive outrage, it stands as a record of shame that, in an enlightened age, no step was taken by a single government to arrest the slaughter of the helpless Christians of Armenia, or to stay the hand of that nineteenth century Nero, Sultan Abdul Hamid, from his sworn purpose of exterminating the Armenian people and thereby ending forever the much-vexed Armenian Question.

Turkish Armenia, the northwest division of Kurdistan, is a great plateau of nearly sixty thousand square miles, bounded on the north by the Russian frontier, by Persia on the east, the plains of Mesopotamia on the west, and Asia Minor on the south. There are in all, at the present time, about four million Armenians on the globe, of whom little more than half are in Turkey, and the rest in Russia, Persia, other Asiatic countries, Europe and America. In Armenia—the name and geographical existence of which are not recognized in Turkey—there are probably six hundred and fifty thousand native Armenians, or one-fourth of the whole number that are scattered throughout the Porte's dominions. The climate is temperate and bracing. Facilities for travel and transportation are exceedingly meagre, and all the methods employed by the natives are unusually primitive. "Valis," or municipal governors, are appointed by the government at Constantinople to administer the laws, and none but Moslems hold official positions. Among the population are found many races, including Turks, Kurds, Russians, Circassians, and Jews, besides native Armenians. Fully one-half the people are Mohammedan.

A KURDISH HOUSE AND ITS INMATES.

The Kurds are tribal and lead a predatory life, dwelling in mountain villages over the entire region. Their number is uncertain, but it is estimated that in the districts of Erzeroom, Van and Bitlis, there are not less than six hundred thousand. Some of these tribes are migratory, like the Bedouins of Syria. Almost all are warlike, and many have degenerated into lawless brigandage. For centuries they have made serfs of the Christians, trampling them under foot at every opportunity, and extending to them no toleration whatsoever. These rude mountain Ishmaelites delight in bloodshed and pillage. A few years ago the Sultan, the better to control them, and with a view to securing for his army an element equal in ferocity and courage to the Russian Cossacks, organized the Kurds into a regular military body with the title of Hamidieh, thus honoring these rough-riding, robber warriors with his own royal name. Their spirit, like that of the wild Arab, the Cossack, or the North American Indian, is one that scarcely brooks the restraints of military discipline. They are always formidably armed, and weapons in the hands of such war-loving races are an incentive to disturbance and outrage. They have long spread universal terror among the Armenians by their cruelty and frightful excesses, but it has been reserved for our own time to witness such an exhibition of barbarism on their part as has filled Europe and America with horror. The Turks, although more civilized, are only one degree less cruel and inhuman than the Kurds. In marked contrast to Kurds and Turks alike, the Armenians are peace-loving, industrious, frugal and kindly. Their nation was converted to the Christian faith in the fourth century, and has remained true to that faith ever since. Their creed and forms of worship are those of the Orthodox Eastern Church ; they believe in the Trinity, and although they cling to many of the ancient forms and symbols, they render no allegiance to Rome. Their native priests or clergy are an earnest, faithful class, and the people themselves hold to their simple faith with an intensity that equals the zeal of the Moslem in supporting Islam. This tenacity of creed, together with the fact that the

BAKING CAKES IN ARMENIA.

Armenians usually prosper everywhere, has been the means of stirring up bitter envy and religious hatred against this peaceable people.

Armenia is a lovely country. It was the first part of the globe to be settled by the human race after the flood, and Mount Ararat, where the ark rested, still rears its lofty crest, seventeen thousand feet in height, and overlooks the same landscape of valley, plain and mountain that greeted the eyes of Noah and his companions when they gazed upon the new-risen earth after the subsidence of the Deluge. In a thousand ways, it has a peculiar claim upon the interest and sympathy of the civilized world. Contemporary with the mighty empires of Assyria, Babylon and Persia, and still later with Rome, it was the birthplace of some of the grandest characters of ancient times. From the earliest days, the nation has worshiped the true God, even though surrounded by idolaters, and its men were famed for bravery and its women for beauty and chastity. Of Prince Ara, one of its rulers, it is related that when urged by the beautiful but licentious Queen Semiramis, of Babylon, to become her husband, he preferred to go to war and lose his life and kingdom, rather than desecrate the sanctity of the Armenian family by such an ungodly union with an idolatrous queen.

MOUNT ARARAT AND "LITTLE ARARAT."

When Christianity dawned upon the earth, its teachers in the first century A. D. found a ready welcome in Armenia, where the Apostles Thaddeus and Bartholomew are said to have preached. Under King Durtad, in the year 302, the Armenians were the first people in the world to accept Christianity as a nation, and the Armenian Church, founded by Gregory, "The Illuminator," has held all the great cardinal truths of the Christian religion throughout the last sixteen centuries, and without a single schism or heresy, or any disrupting theological controversy. Its liturgy was taken from that of St. James of the Church of Jerusalem, and its form of government has been one steady, unchanging line of the Episcopacy, yet without ecclesiastical tyranny. Upon the same patriarchal throne at Etchmiadzin, near Erivan, in Russian Armenia, where once sat Gregory in 302 A. D., now sits the venerable Catholicos Mugurditch Khrimian, the spiritual father of the Armenian people, and well-beloved of all.

KURDISH ROBBERS DISGUISED AS SHEPHERDS.

Mohammedan domination in Armenia dates from the Crusades. Having aided the warriors of the Cross on their outward progress, when the latter were rolled back, discomfited, by the Moslem power, the Armenians were made to feel the bitterness of a revenge such as only a Mohammedan horde could inflict. Their country was overrun and conquered, their property confiscated, even their beloved religion all but suppressed, and their people enslaved. Five centuries relaxed but did not unbind the Moslem bonds. Through many generations these Armenian people have suffered oppression and outrage at Turkey's hands in unresisting silence. Extortions under the name of taxation, gross dishonesty by unpaid officials, and wholesale robbery by the Kurdish chiefs or Agas, together with restricted freedom of worship, and general persecution, made their position almost unbearable.

In 1878, the Berlin Treaty was concluded by the European powers, under which reforms were guaranteed by the Porte in Armenia, whose people were promised security against Kurdish extortions and attacks, and also the fullest religious liberty. Immediately after the Berlin Congress, a treaty of defence was entered into between Turkey and England, and the result has been that the promises made by the Porte to the Berlin Congress, like all others made by the same power, were ignored and broken at every opportunity. From that date, the period of Armenia's worst sufferings was begun. The abuses to which it had before been subjected were now intensified tenfold. Armenians were robbed and beaten, and their stores and houses pillaged at will, their wives and daughters outraged, their cattle and crops carried off, and murder became the common pastime of the Christian-hating Turk. Mohammedan officials ruled in all places of authority, and the word of an Armenian was worthless in a court of justice when opposed to a Moslem. All the laws were distorted for the oppression and degradation of this wretched subject people. At last so loud did the cry of the oppressed become that it again reached the ears of Europe, and the Sultan, being warned, once more, promised to institute reforms in Armenia. He simultaneously registered a vow to exterminate the Armenian people, as subsequent events have shown.

Abdul Hamid's promised reform was inaugurated in September, 1894, by a gigantic and indescribably horrible massacre that has hardly a parallel in history. That it was perpetrated by the Hamidieh—the Sultan's own specially-named troops—is significant of the purpose for which they were organized. The massacre of Sassoon is believed, like all the other great massacres that followed, to have been inspired from the palace at Constantinople, and Zekki Pasha, who commanded on that infamous occasion, was afterward decorated by the Sultan, as were four Kurdish chiefs who had been specially savage and merciless while the carnage was in progress.

THE EDICT OF EXTERMINATION.

From time immemorial, the Armenians have been a rich source of revenue to their Moslem oppressors, who were free to rob, to torture and even to slay them at will. This was the inalienable privilege of the followers of Mohammed in dealing with the "infidel ghiaour." When Europe interfered, and especially when it became evident that such interference, if unchecked, might ultimately lead to the relaxation of Armenia's bonds and possibly even to absolute freedom, the Sublime Porte secretly promulgated a policy as bold and startling as it was inhuman. That policy, which is believed to be the outcome of Abdul Hamid's own brain, is one that stamps that monarch as the supreme savage of the century, and the whole Moslem power as a "barbarian camp," unfit to be tolerated amid civilized nations. Like all Mohammedans, Abdul Hamid's religion is his politics. He regards the life and property of his Christian subjects as his legitimate prey. They are so many dogs, to be whipped or even killed, as the emergency demands; and in the present instance, the Armenians were clearly liable to become a burdensome obstruction to Ottoman Government, and to the peace and serenity of the Sublime Porte. Their tax-paying and tribute-yielding capacity was diminishing, as their numbers and the sympathy of Europe increased. To a true Mussulman, the path of duty was clear. That their importance as a factor in

Turkish affairs might be minimized, they were to be led forth to the slaughter, as other peoples had been in other years, by faithful Sultans. And so the edict of extermination went out from Constantinople, an edict which sealed the fate not only of the people of Sassoon, but of the surplus Christian population of Armenia as a whole. Valis, military commanders and even subordinate officers, in all the principal events that followed, acted under orders from Constantinople. It was a program which, carried out to its fullest extent, contemplated the extinction of the Armenian race within Turkish territory, by the sword, by fire and by starvation. To the Moslem mind, trained to abhorrence of all other religions and urged even by the Koran itself to their subjugation, there was nothing repulsive

ARMENIAN GIRLS SPINNING.

in this, but rather the contrary. How this sanguinary policy was to be put into practice was soon after disclosed.

THE MASSACRE OF SASSOON.

Sassoon is a mountainous province in the southern part of the Armenian plateau, east of Lake Van. Inhabited by Armenians and Kurds, the former are greatly in the majority. There is, however, no intermingling of races. The Kurdish villages are scattered around, being chiefly on the edges of the plateau, while the Armenians dwell in the centre of the province. Industrious and frugal, the Armenians literally supported themselves and the Kurds, and besides paid taxes to the Turkish Government. Of all goods manufactured by the Armenians, the Kurds received their share, or *besh*, as they call it. Every spring, the chiefs or Aghas of the Kurdish tribes, came at the head of their men to collect the tribute from the Armenian villages in sheep, mules, carpets, stockings and implements. The principal taxes which the Armenians pay to the government, are (1) the poll-tax, $2.00 per head, including the new-born male baby; (2) tax on real estate; (3) *Khamtchoori*, namely, five piasters per head of sheep—one-eighth of the value of the sheep; (4) tithe of agricultural products. All these they had honestly paid, but the legitimate taxes had been multiplied tenfold by Kurdish exaction and by the extortions of the valis and minor Turkish officials, each of whom robbed the Armenians at every opportunity. In the Sassoon district, there

A REFUGEE FAMILY FROM SASSOON.

are three Kurdish tribes—the Khanuvdulik, the Busuktzik and the Ousvi—each claiming its own tribute. There are other tribes on the borders of Sassoon—the Pakrantzik, the Baduktzik, the Khiyantzik and the Belektzik, besides many other smaller "ashirets" and all demanded their share. The villages of the Talvoreeg district, richer than most others, paid tribute to seven tribes. Some of the other villages were visited by as many as ten. The wretched Armenians were stripped absolutely bare of everything worth possessing. In 1893, the impoverished Armenians decided to resist further robberies. Early in the spring of that year, the

Kurds came with demands more exorbitant than ever, the chiefs being escorted by a great number of armed men, but they were driven back by the brave villagers. This unsuccessful attack was a new revelation to the Porte. The cry of rebellion was raised and Sassoon was marked for the first act in the drama of Armenian extermination.

In August, 1894, Kurdish and Turkish troops came to Sassoon. The Kurds had been newly armed with Martini rifles. Zekki Pasha, who had come from Erzingan, read the Sultan's order for the attack, and then urged the soldiers to loyal obedience to their Imperial master. It is said that on the last day of August, the anniversary of Abdul Hamid's accession to the throne, the soldiers were specially urged to distinguish themselves in making it the day of greatest slaughter. On that day the commander wore the edict of the Sultan on his breast. Kurds began the butchery by attacking the sleeping villagers at night and slaying men, women and children. For twenty-three days this horrible work of slaughter lasted. No pen can adequately describe the diabolical ferocity of the prolonged massacre. Some of the Kurds afterward boasted of killing a hundred Christians apiece. At one village, Galogozan, many young men were tied hand and foot, laid in a row, covered with brushwood and burned alive. Others were seized and hacked to death piecemeal. At another village, a priest and several leading men were captured and promised release if they would tell where others had fled; and, after telling, all but the priest were killed. A chain was put around his neck and pulled from opposite sides until he was several times choked and revived, after which bayonets were planted upright and he was raised in the air and dropped upon them. The men of one village, when fleeing, took the women and children, some five hundred in number, and placed them in a ravine where soldiers found them and butchered them. Little children were cut in two and mutilated. Women were subjected to fearful agonies, ending in death. A newly wedded couple fled to a hilltop; soldiers followed and offered them their lives if they would accept Islam, but they preferred to die bravely professing Christ. On Mount Andoke, south of Moosh, about a thousand persons sought refuge. The Kurds attacked them, but for days were repulsed. Then Turkish soldiers directed the fire of their cannon on them. Finally the ammunition of the fugitives was exhausted, and the troops succeeded in reaching the summit unopposed and butchered them to a man. In the Talvoreeg district, several thousand Armenians were left in a small

A KURDISH CHIEF.

plain. When surrounded by Turks and Kurds they appealed to heaven for deliverance, but were quickly dispatched with rifles, bayonets and swords. The plain was a veritable shambles.

No accurate estimate of the number slain in the first massacre has been made. Forty villages were totally destroyed and the loss of life is believed to have been from ten to fifteen thousand. Efforts were made to conceal the real extent of the carnage, but the "blood-bath of Sassoon" has now passed, into history and cannot be forgotten.

ARMENIANS KILLED IN THE STREETS.

Some of the incidents connected with this widespread slaughter in the Talvoreeg district, between Moosh and Diarbekr, were of a nature to strike the civilized world with horror. It is said that no respect was shown to age or sex; men, women and infants were treated alike; the women being subjected to greater outrage before being slain. In one place, about two hundred weeping women knelt before the Turkish commander, pleading for life, but the brutal officer ordered them to be served like the others. One letter describing the massacre said: "Some sixty young brides and other attractive girls were crowded into a little church where, after being assaulted, they were slaughtered and a stream of human blood flowed from the church door." To some women in one village the proposition was made that they might be spared, if they denied their faith. "Why should we deny Christ?" they said, and pointing to the dead bodies of their husbands and brothers before them, they nobly answered, "We are no better than they; kill us too"—and so they died. A priest was taken to the roof of his church and hacked to pieces; young men were placed among wood saturated with kerosene and set on fire. After the massacre, and when the terrified survivors had fled, there was general looting by the Hamidieh Kurds. They stripped the houses bare, then piled the dead into them and fired the whole, intending, as far as possible, to cover up the evidences of their dreadful crime.

So great was the indignation in Europe over the Sassoon slaughter, that a Consular Commission of Inquiry was demanded for the purpose of investigation. After a long investigation, a report was made which was only a partial confirmation of the truth. From the outset everything was against the Commission, and especially against the efforts of the European delegates. In Van, Bitlis and elsewhere, witnesses were arrested and intimidated by the government.

Comparative order prevailed for a time during the period of the Commission's sitting, but it was a delusive calm. Its work completed (early in 1895), promises of new administrative reforms were made by the Porte, but almost as soon as the field was again clear, the massacres recommenced with redoubled vigor. The Kurdish Hamidieh were again brought into requisition, and the Mohammedan populace in all the large cities of Asia Minor were deliberately inflamed against the Armenians by circulating lying rumors of intended attacks on the mosques. Soon there was an outbreak at Constantinople in which nearly two hundred Armenians were killed by the "Softas," or Mohammedan students, and the police. This was followed by a terrific outburst of fanaticism all over the Sultan's empire, and by such scenes of massacre as have not been paralleled since mediæval times. Throughout all the vilayets of Armenia ran the red tide of murder. Hundreds of villages were swept away, and their inhabitants either slain or exiled. In this work of destruction the Kurds played the most prominent part, but soldiers and Turkish civilians did their full share. The object was to destroy everything so effectually that the Armenians would have no means of living, and would have to choose between death and Islam. Their cattle and all movable goods were carried off, and everything else destroyed. In some villages even the clothing was taken from the backs of the wearers, and they were left literally naked. Abdul Hamid's government was completing its diabolical work by reducing the population and then confiscating property under the pretended forms of martial law, and by forcing the starving Armenians to apostatize to save their lives. In some places the poor wretches yielded to the pressure, but the greater number held out staunchly for their faith, many dying rather than surrender their Christianity.

THE LATER MASSACRES.

In the absence of accurate data it is, of course, impossible to give a reliable estimate of the multitudes of Christian Armenians who perished in the great slaughter that followed Sassoon. The figures given below are approximate, and as they are compiled from Turkish sources, may be regarded as rather under than above the mark. According to Turkish calculations, the number of those who were in a condition of starvation in February, 1896, was one-half the agricultural population of the vilayets (or districts governed by a Vali or Pasha) of Anatolia, (the Turkish name for Armenia) being about 275,000 souls, of whom two-thirds were women and children. The figures below present a conservative view of the results of the Sultan's policy of extermination during the first sixteen months:

Name of Town.	Date of Massacre.	No. Killed.	By Whom Done.
Sassoon, ...	Aug.–Sept.,	10,000	Kurds and Turks.
Constantinople, . . .	September 30,	172	Police and Softas.
Ak-Hissar, ...	October 9,	45	Moslem villagers.
Trebizond, ...	October 8,	1,100	Soldiers, Lazes, Turks.
Baiburt, ...	October 13,	1,000	Lazes and Turks.
Gumushane, ...	October 11,	550	
Erzingjan, ...	October 21,	1,900	Soldiers and Turks.
Bitlis, ...	October 25,	1,200	Soldiers, Kurds and Turks.
Harpoot, ...	November 11,	1,000	Soldiers, Kurds and Turks.
Sivas, ...	November 12,	1,200	Soldiers and Turks.
Palu, ...	October 25,	1,200	Soldiers, Kurds and Turks.
Diarbekr, ...	October 25,	2,500	Soldiers, Kurds and Turks.
Albistan, ...	October	300	
Erzeroum, ...	October 30,	1,200	Soldiers and Turks.
Ourfa, ...	November 3,	400	
Kara-Hissar, ...	October 25,	500	Circassians and Turks.
Malatia, ...	November 6,	250	
Marash, ...	November 18,	1,000	Soldiers and Turks.
Aintab, ...	November 15,	. .	No details.
Gurun, ...	November 10,	3,000	Kurds and Turks.
Arabkir, ...	November 6,	2,000	Kurds and Turks.
Argana, ...	. . .	. .	No details.
Severek, ...	. . .	. .	No details.
Tokat, ...	. . .	. .	No details.
Amasia, ...	. . .	. .	No details.
Marsovan, ...	November 15,	125	Turks.
Kaesarea, ...	November 30,	1,000	Circassians and Turks
Gemerek, ...	. . .	500	
Egin, ...	. . .	. .	No details.
Zileh, ...	. . .	. .	No details.
Sefert, ...	. . .	. .	No details.
Khnous, ...	. . .	300	
Boulinek, ...	. . .	400	

The Turks estimated the "reductions" made by Abdul Hamid's slaughter policy as follows:

Armenian population in larger towns,	177,700
Armenian population in villages,	538,500
Number killed in towns (estimated),	20,000
Number of Armenian villages (about),	3,300
Number of villages destroyed (estimated),	2,500
Number killed in villages,	No accurate data.
Number reduced to starvation in towns (estimated),	75,000
Number reduced to starvation in villages (estimated),	360,000

That these figures fall far short of the actual result of this series of wholesale assassinations is not disputed. Consular officials, missionaries and the few correspondents unite in declaring that the total "reduction" of population by this sanguinary method has in every instance been underestimated. But cold figures and dry statistics can tell nothing of the unspeakable horror of those days of blood, and of the tortures suffered by the Christian population of the cities of Armenia at the hands of their murderers. It is mainly from letters received by Armenians in the United States from surviving relatives at home, that the true story of those dreadful months has been gleaned. At Trebizond, where eleven hundred perished, "only a few Turks were killed," says a letter from a fugitive. "Like a clap of thunder in a clear sky, the thing began about 11 a. m., October 8. Unsuspecting people walking along the streets were shot ruthlessly down. Men standing or sitting quietly at their shop doors were instantly dropped with a bullet through their heads or hearts. The aim was deadly, and I have heard of no wounded men. Some were slashed with swords until life was extinct. They passed through the quarters where only old men, women and younger children remained, killing the men and large boys, generally permitting the women and younger children to live. For five hours this horrid work of inhuman butchery went on, the cracking of musketry, sometimes like a volley from a platoon of soldiers, but more often single shots from near and distant points, the crashing in of doors, and the thud, thud of sword blows sounding on the ears. Then the sound of musketry died away and the work of looting

TREBIZOND, WHERE A MASSACRE TOOK PLACE.

began. Every shop of an Armenian in the market was gutted, and the victors in this cowardly and brutal war glutted themselves with the spoils. For hours, bales of broadcloth, cotton goods and every conceivable kind of merchandise passed along without molestation to the houses of the spoilers. The intention evidently was to impoverish and as near as possible to blot out the Armenians of the town. To any found with arms, no quarter was given. Some were offered life, if they

ARMENIANS HELD PRISONERS AFTER THE TREBIZOND MASSACRE.

renounced Christ and accepted Islam; but large numbers were shot down without any proffer of this kind. One poor fellow when called on to surrender, thought he was called on to give up his religion, and when he refused, he was hacked to pieces in the presence of his wife and children.

"Over five hundred Christian Armenians were slaughtered in the neighboring villages. Untold horrors are implied in this brief statement. Many Armenian women vanished, having been either murdered or kidnapped; most of the

Armenian houses were burned to the ground, the survivors being driven like wild beasts to the hills and woods.

"Equally sad was the fate of the Christians of Baiburt, whose tragic taking off is related in a letter addressed by the survivors to the Armenian Patriarch at Constantinople. After giving a partial list of the slain, the writers state: "When the massacres and plundering began, on account of the prevailing terror and insecurity, the people were compelled to close all the churches, shops, and schools from October 13 to 26, and take refuge in the houses. Letters were sent from our prelate to the commandant of the Fourth Army Corps, at Erzeroum, and to the Armenian prelate at Erzeroum asking assistance, but all our prayers remained unanswered. After the massacres the Turks advised us indirectly that the order was secretly given from the Imperial Palace and was irrevocable! It was on Saturday, October 22, that the fatal hour struck. The frantic Turkish mob, assisted by regular troops, suddenly fell upon the innocent and unarmed Armenians. The bloody work began at 4 o'clock a. m., and lasted until 12 o'clock in the evening (Turkish time). Besides murdering our people, the mob plundered and fired the Armenian dwellings and stores, taking care that the Greeks should not be molested. On that frightful day the Armenian community of Baiburt was almost annihilated. Strong men, youths and women and even babies in their cradles, and unborn children in the wombs of their mothers, were butchered. Infants were stuck on bayonets and exposed to the view of their helpless and frantic mothers. Young brides and girls were subjected to a fate that need not be described. No resistance was possible on the part of the Armenians. All the native teachers, with a single exception, were murdered with most cruel tortures. Baiburt became a slaughter-house. Torrents of blood began to flow. The streets and the bazars were filled with dead bodies. On the following day the Turks did all in their power to conceal the bodies of those who had been pierced by bayonets. Similar scenes were enacted in the surrounding villages."

MGR. IZMIRLIAN.
Armenian Patriarch at Constantinople.

The Harpoot massacre was another butchery carried out under orders. This was one of the leading stations of the American Mission. Sixty Christians fled to a church in the vain hope that its walls would furnish them a shelter against those who were crying for the blood of the Armenians. They were permitted for a time to believe themselves secure, but suddenly the church was surrounded by a great number of Kurds. The doors were then blown in, and the Christians thought that they would be massacred within the sacred structure. They were not. Their captors took them one at a time outside the church, and there, heedless of the cries for mercy from women and children, killed them, either by shooting or stabbing them. The first victim was the Protestant pastor of the church, who, as he was dragged out, bade the others, if they had to die, to die as Christians. He met his death like a martyr. Some of the refugees, in a very agony of terror, offered to abjure their faith and accept Islamism, thinking thus to save their lives. The offers availed them nothing, for their insatiable enemies, after accepting them, dragged the converts out and killed them one by one. The Armenian Church has been turned into a mosque, and the Protestant Church into a stable.

A missionary tells the story of the desolation of Harpoot as it was related to him by an eye-witness who saw the Christian quarter in flames and the houses of the American Mission burning. He came on to Malatia (the ancient Melitene), and found not a house in the Christian quarter standing. In a khan there were about twenty wounded men, the sole survivors of a caravan of 200 who had been traveling to Harpoot from Northern Syria and whose members had nearly all been slain by the Kurdish bands. There were 150 dead bodies lying in the road. At Marash (another mission station of the American Board), the same witness, days after the massacre, counted eighty-seven dead Armenians in one spot, and there were hundreds of bodies strewn around in the near neighborhood. In the villages on the plains near Harpoot, each containing from fifty to 1000 houses, the evidences of slaughter were sickeningly abundant. The Kurdish butchers had slain fully half the population. The door of a house would be burst open, a volley fired upon the shuddering inmates, while those who rushed out were caught and killed in the fields. Then the houses were plundered, fired and left blazing. This was the fate of thousands of Christian homes.

It is proved beyond doubt that the massacre at Erzingjan started in the office of the Vali or local governor, where an Armenian priest of Tevnik was shot down by Turkish assassins. Then followed a horrible carnage, during which over one thousand Christians were slaughtered. After the butchery, the dead victims were dragged by neck and heels into the cemetery and cast into a long, deep trench, not unlike the death pit of Galogozan—the murdered fathers, mothers and sweet, innocent babes, all calm and peaceful in the sleep of death, flung down like carrion. Nothing more horrible or pathetic could be imagined than that scene at the cemetery two days after the massacre. The survivors dared not even express their grief.

But the climax had not yet been reached; the appetite of the Moslems for Christian blood had merely been whetted, not satiated. Other and equally terrible butcheries followed at Karahissar, Arabkir, Ouloupinar, Palu, Mardin, Sivas, and Tchoukmerzen, where Kurds and Turks perpetrated wholesale murders and swept large districts desolate. The villages round about Erzeroum were almost depopulated, the orders for the slaughter of the Christians, as the Moslem troops admit, having come from Constantinople. At Sivas the massacre was terrible, and a like horror occurred at Marash. The ungovernable fury of the Turks spared neither age nor sex, and the brutalities practiced upon women and children may not be described. In the Erzeroum massacre fully twelve hundred perished, including women, many victims being mutilated. Bodies of little children, dead and mutilated, were found in the fields after the slaughter had ended. Large numbers of the victims of these atrocities died the death of martyrs. They fell in the Moslem war for the extermination of the religion of Jesus in Asia Minor.

At Diarbekr, where the victims were numbered by thousands, there was abundant evidence that the massacre was premeditated. It was claimed that the Armenians had attacked a Moslem mosque, whereas the facts, as afterward disclosed, showed the Kurds and Turks to have been the sole and intentional aggressors. The massacre began on Friday, and continued on Saturday and Sunday with insatiable ferocity.

CHILD-VICTIMS OF THE ERZEROUM SLAUGHTER.

Meanwhile, the story of what was taking place in the villages and hamlets of the different districts had not reached the public ear. When it came, it disclosed a tale of suffering and savagism that has scarcely a parallel. Many hundreds of villages were literally swept out of existence. The story of one is the story of all: the Kurds, directed from higher sources, swooping down, rounding up the cattle, slaying the strong men, outraging and abducting the women, and killing even the children, concluding the satanic work by burning everything that would consume. In many places the Kurdish troops came equipped with empty sacks strapped to their saddles for the purpose of carrying off the plunder. The Kurdish chiefs openly declared that they were ordered to slay the Christians and take the plunder for their pay.

An illustration of the Turkish method of extermination is found in the case of the village of Hoh, in the Sandjak district. At first the "aghas" (or local magistrates) promised to protect the Christians, but when they saw villages burning in every direction, they refused to keep their word. All the Christians were told that, under the pain of death, they must accept Islam. They were assembled at the mosque, and there eighty young men were picked out and led outside the village—for slaughter. Eight escaped, sixty-two were killed, and ten wounded. The young women of the village were taken to Turkish harems. In and around the villages of Kenerik, Moorenek and Rusenik, and the town of Mardin, fourteen native preachers were killed, several being hideously tortured before they were dispatched. During one of the days of massacre at Kæsarea, an attack was made on the public baths. Six naked Armenian women were dragged forth and bayoneted. Young girls were drawn through the streets by the hair and the feet. Eight of the villages near Van are totally depopulated, all their people slain or fled, except the young women who have been seized and taken to Kurdish harems. In Van province nearly 200 villages have been partially destroyed. Eleven villages around Harpoot were forced to accept Islam unconditionally or die. The wretched people were then set to killing their fellow Armenians, to prove the genuineness of their conversion. Such horrible tortures as flaying alive, cutting to pieces by swords, tearing out the eyes, branding on the body with red-hot irons, and even tearing out the entrails, filling up the cavity with gunpowder and exploding it—these were among the simplest of the diabolical measures adopted by the Sultan's officials and his soldiery in dealing with his Christian Armenian subjects. Women torn from their homes and outraged, and hundreds of young girls forcibly carried off, fiendishly used and wantonly slain, and other horrors unnamable, were some of the methods employed in upholding the glory of Islam.

REFUGEES ON THE TURKO-PERSIAN FRONTIER.

THE SUFFERING AND DESTITUTION.

These persecutions and wholesale massacres, together with the general destruction of property, reduced the Armenian survivors to a condition of utter destitution. From the ruined villages, the now homeless women and children flocked to the cities and towns, while the remnant of the male Armenians were fain to hide in the mountains. There was a condition of universal suffering which the Turkish Government seemed resolved should have the effect of finishing the work of extermination so well begun—death from starvation and exposure would soon claim the survivors. Thousands had fled to the forests and the mountains; the survivors of Sassoon were living in caves, and subsisting upon berries and roots until they became livid like corpses. "Hunger-bread," a horrible compost of chopped straw and roots, pounded together and baked, helped to keep the life in their emaciated bodies. The babes and the weak women could not survive such a diet, and they were quickly perishing when the Christian missionaries came, like angels of blessing, with help, in the shape of food and clothing. Many had already died of hunger and cold, and all were more or less naked. Meanwhile Van was inundated by refugees, and also the cities along the Persian border; while the interior cities were all filled with crowds of destitute who had flocked thither from the ruined villages. All Armenia was reduced to a race of naked beggars. Thousands of families, lately prosperous, were now destitute, their bread-winners slain, their homes in ashes, and even their little stores of food destroyed, so that they might starve the quicker! Yet had they, even at this juncture, been disposed to yield, as some did, to the Turkish offer to abjure Christ and turn Mohammedan, persecution would have ceased and they might again have been prosperous, with their property restored. But the Armenians, although a simple people, have the strong, sturdy character of which martyrs are made, and to their honor be it recorded that in a majority of instances the offer was spurned. They would rather die than become apostates to the faith of their fathers!

HUNGER BREAD FROM BITLIS.

Very striking is the testimony of some of our most esteemed missionaries to the Christian fidelity of the Armenian people. Probably the best known and most experienced of all the Americans who have served in the missionary field in Asia Minor is Rev. Cyrus Hamlin, D. D., the venerable founder of Roberts College, Constantinople. Dr. Hamlin, who is now in the United States, has a life-long acquaintance with the Armenian question in its various phases and is a strong champion of the right of this oldest Christian nation on earth to be permitted to live and worship in the faith of their fathers. Conversing recently on the subject of Armenia's sufferings Dr. Hamlin said to the writer: "The condition of affairs in that country has not been exaggerated in the printed reports. I have lately finished reading some two hundred letters from missionaries, a very large part of them dealing with the oppressions and sufferings of the Armenians, which were of a most frightful character. The whole civilized Christian world should help these people—they should be saved from death. They can look in no other direction for help, for there is no sympathy and assistance to be had from Turkey. Indeed, the policy of the Sultan's government is apparently dictated by a desire to efface the Armenian people altogether—at least those of them who will not accept

ARMENIAN WOMEN MAKING BREAD.

Mohammed. When you talk sympathizingly about these people, a Turk will say in surprise: 'Why do you speak in behalf of such worthless trash and try to save them? They can save themselves—all they need to do is to accept Islam and then they are safe and out of trouble.' A Turk regards it as strange that an Armenian should refuse to purchase his life at the cost of his faith; but there are some among them who take a different view. Some of the Turkish soldiers, who shared in the terrible atrocities lately perpetrated on the Armenian Christians, have been stricken by remorse afterward. One soldier, who had borne his part in several horrible butcheries of women and children, was so troubled that he could not sleep. He had visions of his victims that ultimately drove him insane.

"Mrs. Knapp, a missionary at Bitlis, related a remarkable incident. A soldier, who had aided in the ruthless massacres of the helpless ones, was terribly tormented by conscience. To his wife he said: 'There was one thing about those women and their children that I do not understand and I want you to ask the wives of the "ghiaour" (Christians) about it. It was very strange. The women were offered their lives if they would only say: "There is but one God and Mohammed is His prophet," but they would not. They all died in terrible tortures, calling on "Hissos Nazareetsees." That is what I do not understand. Now, I wonder who this Hissos Nazareetsees is, whose very name made these

A RELIEF COMMISSIONER PASSING MOUNT ARARAT.

women so brave that, with their little children, they could die. That is what troubles me greatly.'

"The good missionary explained to the Moslem wife, who, in turn, told her husband, that the name was that of the worshipful Jesus of Nazareth, Saviour of the world, whom the Christians serve."

THE RELIEF MOVEMENT.

Appeals representing the condition of the Armenian people as deplorable beyond description, touched sympathetic hearts in Europe and America and a general movement for their relief was begun. This, however, did not suit the purposes of the Turkish Government, which declared its entire ability to take care of its own, and even denied the palpable fact of universal Armenian destitution, as it had previously denied the perpetration of the massacres. In England a fund was raised, under the auspices of the Duke of Westminster, and distributed through Consular officials and American missionaries, the Armenians resident in Europe and America contributing toward it. Dr. Louis Klopsch, of New York, dispatched a commissioner to Van to ascertain the exact facts concerning the need of the people and to organize a Relief Committee of American missionaries. Its Commissioner, William Willard Howard, was not permitted to cross the Turko-Persian frontier, being excluded by Turkey. He made a number of attempts, at the risk of his life, to push his way through. Passing near Ararat, in a lumbering stage, he was attacked by Kurds. Again, on a second attempt, the horse he rode was shot and he himself narrowly escaped. In still another effort to cross the frontier he had a regular pitched battle with Kurds, a number of whom, disguised as shepherds, were lying in wait for travelers whom they might rob and slay unhindered, the whole country being at war. Many, besides Armenians, have met their fate at the hands of those Kurdish murderers. Mr. Howard took the caravan route through Russia and Persia, *via* Batoum, Tiflis, Erivan and Khoi, and so across to Van, keeping close to the Turko-Persian border for a considerable part of the journey. At the frontier he was driven back by the Turkish officials and, menaced by their Kurdish allies, he reluctantly gave up the effort to enter Van. Mr. Howard's failure, however, did not deter the *Christian Herald* from carrying out its humane project, for, with the co-operation of the missionaries of the American Board in Van, it organized a most successful relief work, partly industrial and partly charitable, under the active personal supervision of Dr. Grace N. Kimball, a medical missionary. Through these means several thousands of the needy were fed and supported in Van daily. Other relief stations were opened by the same journal at Erzeroum, Erzingjan, Harpoot, Diarbekr, Mardin, Gemarek, Aintab, Sivas, Arabkir and several other points which had been the scenes of massacre and where the suffering was most acute. On these relief stations a fund of nearly $30,000 was expended. An effort was made by the American Red Cross to obtain permission to visit Armenia and distribute relief, but its application met a decided

AN ARMENIAN BEGGAR IN VAN.

refusal from the Sultan's government, although, at the time, the necessities of the Armenian people were greater than ever and hundreds were perishing of cold and starvation.

In the noble relief work that was being conducted amid so many perils, one figure stands out boldly, that of a woman, delicately reared and highly cultured, yet brave to face even death in the Lord's work, to which she had dedicated herself. Dr. Grace N. Kimball will long be remembered as the heroine of Van, whose courage and nobility of soul were the means of saving probably thousands of precious lives. As the first wave of persecution and slaughter receded, and the fugitives were flocking to Van, sick, indigent and nearly naked, Miss Kimball gathered what funds she could and quietly and without any preliminary flare of trumpets, began a systematic work of relief, which had already achieved excellent results before the startling series of massacres began in the fall of 1895. There were many times when Dr. Kimball and her associates were imperiled in consequence of their relief work, as the Turks resented all sympathy with the Armenians or the extension of any aid that would prolong their lives. But all stood bravely at their posts. So with the American missionaries at Harpoot and Marash (where the mission buildings were burned down after being looted) and at every other point throughout Armenia. Although warned by United States Minister Terrell at Constantinople to leave and, with their wives and children, go to the coast for safety, the brave missionaries clung to their posts, preferring to stay by and help the victims of persecution and if need be even to die with them, rather than leave them to the cruel mercies of the Turks. And they were sorely needed, for every day increased the suffering. Before October, 1895, a large number of Armenians had actually died of hunger. Those who saved themselves by flight reached safety in rags, many with only a single garment to protect them against inclement weather. United States Consul Graves, writing from Talvoreeg, thus described the condition of these people: "Bread they have not tasted for months, and curdled milk they only dream of, living, as they do, upon greens and the leaves of trees. There are two varieties of greens which are preferred, but these are disappearing, as they wither at this season. Living on such food, they become sickly; their skin has turned yellow, their strength is gone, their bodies

DR. GRACE N. KIMBALL.
"The Heroine of Van."

are swollen, and fever is rife among them." A touching picture of the gratitude of the sufferers on receiving relief from the missionaries, is contained in a recent letter from Van: "Men and women," the writer says, "come to us, their eyes streaming with tears of gratitude, and clasp the missionaries' knees, and even prostrate themselves, kissing the hands and feet in their gratitude. Many missionaries even have no shelter and are compelled to sleep on the naked earth, while attending to the relief work." In all the larger cities of Armenia—Van, Aintab,

DESTITUTE ARMENIANS BEFORE MISS KIMBALL'S RELIEF STATION AT VAN.

Bitlis, Erzeroum and Trebizond, the streets are filled with pitiful-looking crowds of fugitives, haggard and emaciated. They come from the country districts, which the massacres have, in many places, swept as bare as a desert. In a few of the cities, little bands of American missionaries, aided by the Consular officials, stand between thousands and death. Hundreds of Christian churches have been desecrated by Kurds and Turks, their fonts and altars befouled with offal, their sacred vessels stolen and the buildings either burned or transformed into stables or mosques. The Turkish jails are full of prisoners, all Armenians, arrested on the most frivolous pretexts, or on none at all, the general charge being rebellion. Such is the horribly unsanitary condition of those jails (as at Trebizond and Erzeroum), that few will come out alive. Many have already died from the effects of their imprisonment.

THE ARMENIAN REVOLUTIONISTS.

It has been invariably asserted by the Sultan's Government that the Armenian troubles were the outcome of a deep and widespread revolutionary movement, and that the Turks themselves, rather than the Armenians, were entitled to commiseration. These revolutionists, who were controlled by a patriotic Armenian society known as the Huntchaugists, were directed by a governing power outside of Turkey. Their emissaries were everywhere, and they were constantly fomenting disturbance between Turk and Armenian. They had imported arms and money into Turkey and it was at their instigation that the rebellion broke out in such formidable force at Sassoon, which the Turkish army found much difficulty in quelling. It was due to the influence of the Huntchaugists too, and under the inspiration of their example that the Armenians in other places had arisen against the kind and beneficent government of Abdul Hamid. Indeed, so formidable was this insurrectionary movement that the Porte had been compelled to use force in disarming the rebellious populace of the large cities, and the latter in several instances had so stubbornly resisted that blood had flowed, and many innocent and inoffensive Moslems had perished at the hands of the desperate Armenian rebels. Incidentally, some of the latter were doubtless also slain; but this fate only happened to them when in open rebellion. When the good-hearted Queen Victoria wrote a letter to the Sultan expressing regret over the disorders in Asia Minor, Abdul Hamid explained that the troubles in Anatolia (the name Armenia having no geographical existence in Turkey), had been precipitated by the Armenians themselves, that the printed reports in the British press were wilful exaggerations and that far from the Armenians being the greatest sufferers, a majority of the victims were Turks! He professed regret that Her Majesty should believe any further disorders possible, in view of the reforms he had decided to inaugurate in the disturbed districts.

This "ostrich" policy of denying what is obvious to the whole world, is characteristic of the Sublime Porte. With the facility for intrigue, distortion and falsification, which peculiarly belongs to the Oriental, Abdul Hamid and his ministers, have endeavored by constant prevarication, to hoodwink Europe as to the real status of the Armenian case. But the Turk's pose as a martyr and a saint is an ineffective one, and the mask is easily penetrated.

One of the most mendacious statements circulated by the Ottoman Government was the charge that the massacres were deliberately invited and provoked by the Huntchaugists, that they planned the disturbances, knowing that the result would be death to thousands of their fellow-countrymen and women, yet satisfied even at such a fearful cost to excite the sympathy and provoke the interference of Russia or some other great power. This charge has been emphatically denied, and is so wholly brutal and out of harmony with the Armenian character, as to be utterly unworthy of belief.

It is undoubtedly true that, when pressed to the last ditch, the Armenians at different places made a desperate stand for their lives. These were gleams of heroism amid the massacres that lighted up the darkness as the sunshine glints through the storm-clouds. At Zeitoun, a fortified town of Armenia, the Christian townspeople took arms and made a brave resistance. They mustered in force, captured the citadel and turned its guns on the dismayed Turks, having first provided for the safety of their wives and children. Bravely they held out for weeks, and a strong force under Mustafa Pasha failed to dislodge them or recapture the town. Resistance was also encountered by the Turks and Kurds at the hands of the Christians of Diarbekr, and many fell under the Armenian attacks, although the latter were finally overpowered and massacred. Some of the villages too opposed a brave resistance. But in the end, the story was the old familiar one of overwhelming forces and cruel butchery.

ATTITUDE OF THE EUROPEAN POWERS.

While the great crime against the Armenian people was being enacted, and even while the red tide of massacre was at the flood, Europe looked on with apparent indifference. The leading powers of Christian Europe—Germany, Russia, England, France and Italy—had their magnificent fleets riding at anchor within reach of Constantinople, and a single resolute remonstrance would have been heeded by the Turks, and might have saved many lives. But the word remained unspoken, the cannon lay silent, while a Christian nation was being exterminated. The six great powers were dead-locked in hopeless impotence. Russia, it was believed, would have consented to occupy Armenia and to compel a cessation of the massacres, but England would not yield assent. Germany, too, had its jealousies, and Italy, France and Austria were each so intent on watching the movement of the other powers that none of the three cared to bestir themselves. The United States was represented in the Levant by several war vessels, for the purpose of affording a certain assurance of protection to American interests and the American missionaries; but our naval demonstration was insufficient to save from destruction the American Board Mission buildings at Harpoot and Marash, which were burned by Turkish mobs during the riots and massacres in those cities.

But if governments were inactive, Christendom was active, and leading men, in both Europe and America, were loud in their denunciation of the Sultan and his bloodthirsty policy. Lord Salisbury, the English Premier, speaking on a public occasion, at a time when the patience of Europe had apparently been well-nigh exhausted, said: "Above all treaties, all combinations of the powers, in the nature of things, is Providence. God, if you please to put it so, has determined that persistent and constant misgovernment must lead the government which follows it to its doom. The Sultan is not exempt any more than any other potentate from the law that injustice will bring the highest one on earth to ruin." According to latest advices, the English Government was still depending on Providence to save the Armenians. It had seemingly forgotten its own sacred pledge to secure to that afflicted people the right to free worship and the several reforms conceded under the Berlin Treaty. Tenfold stronger was the emphasis employed by Mr. Gladstone in a public utterance on the massacres. That eminent statesman, replying to a delegate of Armenians, said:

"We may ransack the annals of the world; but I know not what research can furnish us with so portentous an example of the fiendish misuse of the powers established by God for the punishment of evil-doers, and for the encouragement of them that do well. No government ever has so sinned; none has so proved itself incorrigible in sin, or, which is the same, so impotent for reformation. I have lived to see the empire of Turkey in Europe reduced to less than one-half of what it was when I was born, and why? Simply because of its misdeeds—a great record written by the hand of Almighty God, in whom the Turk, as a Mohammedan, believes, and believes firmly—written by the hand of Almighty God against injustice, against lust, against the most abominable cruelty. Such a government as that which can countenance and cover the perpetration of such outrages is a disgrace to Mahomet, the Prophet whom it professes to follow, it is a disgrace to civilization at large, and it is a curse to mankind."

HOPE DAWNS FOR ARMENIA.

On January 23, 1896, a new and totally unexpected development of the Armenian question occurred, which took Europe by surprise. Throughout the Armenian troubles, and especially when the censure of Europe was strongest against the Porte, Russia maintained an attitude of friendly tolerance toward the government of Abdul Hamid. When the other powers proposed that their respective governments should have the privilege of an extra guardship in the Dardanelles, Russia, through its ambassador at Constantinople, waived any claim to such concession by the Sultan. It was due to the influence of Russia also that an attempt by England to make a naval demonstration before Constantinople was abandoned. Germany and France, while ostensibly on friendly terms with the Czar's government, really stood in constant apprehension of some bold, defiant stroke by Russia, that might incidentally either strengthen or shatter their friendly relations, while adjusting matters with Turkey to its own satisfaction. England, suspiciously standing aloof from all alliances, yet pretending in turn to friendship for Russia, France, Italy and Germany, occupied a unique position. As the special patron of Turkey, she was to a large extent responsible for the series of frightful massacres which had disgraced Europe, and made of Asia Minor a region of death and desolation. Yet England, up to the very last, took no step to stop the butchery, but satisfied her conscience with mild official and unofficial remonstrances, and ineffective political manœuvring. At the height of the troubles, it was intimated to England that Russia stood ready to occupy Armenia with an armed force, and to undertake the pacification of that country, but would do so only with the consent of all the powers. To this proposition England turned a deaf ear. Never would she consent to such a scheme; her jealousy of Russia's growing influence in the East forbade it altogether. The Armenians, if saved at all, must be saved by some other power. Rather than see their country come under Muscovite domination, even temporarily, Lord Salisbury's humane policy would prefer the continuance of the massacres as the lesser evil.

Suddenly came the news—late in January, as already stated—that a treaty or compact had been concluded between Russia and Turkey, for offensive and defensive purposes, under which Russia agreed to defend the Dardanelles, in the event of war against either country, and also to restore order in Armenia. The treaty, while it guaranteed the integrity of the Ottoman Empire, also made the Czar the master of the Dardanelles. France, by a secret understanding with Russia, consented to the treaty and agreed to support the Czar's government throughout. Germany and Austria were also supposed to be consenting and interested powers, Italy and England being ignored. Thus, by a single coup, M. Nelidoff, the Russian ambassador to Constantinople, won for his royal master a double diplomatic triumph: securing the pacification of Armenia and the cessation of the massacres without involving Europe in a general war over the dismemberment of Turkey, and grasping for the Czar the splendid prize for which Russia has long hungered: the sovereignty of the Dardanelles.

With the new order of things, and with Sultan Abdul Hamid as a treaty vassal of the Czar, there comes a gleam of hope to Armenia, the hope of peace and brighter days to come. Many years must elapse, however, ere the "blood-bath of Sassoon," the death-pit of Galogozan and the other dark memories of the terrible period of 1894–96 be forgotten, even by those who were children when these events occurred. But the fathers and mothers of Armenia, who have shared in those sufferings, will carry the recollection with them to the grave.

ACTION OF OUR GOVERNMENT.

American sympathy for Armenia's sufferings took a more direct and practical form than that of any of the European countries. Clara Barton courageously proceeded to Constantinople, confident that the Sultan could be persuaded to relax his opposition to the Red Cross entering Armenia on its work of relief. Not only did the American people send generous contributions of money to feed the starving refugees, but the press of the nation, not standing in awe of any alliances, was unanimously outspoken in its strong condemnation of the barbarous policy of the Porte. A concurrent resolution was introduced in Congress, looking to the amplest protection for American citizens in Turkey, and directing that our government ask the European powers to act promptly for the prevention of further bloodshed, and a repetition of the massacres. From all parts of the Union, the President and Congress received, almost daily, a multitude of letters, petitions and memorials, urging that the time had arrived when the United States, as a Christian nation, should place on record its abhorrence and condemnation of the bloodthirsty and fanatical Ottoman policy in Asia Minor. Returning missionaries, many of them coming from places that had been the scenes of massacre, confirmed the stories of outrage and slaughter and deepened the impression already made by the recital of Armenia's woes. From the pulpits of all Christian denominations came thunders of eloquent denunciation against the Turks. England, whose fleet had been stationed off Constantinople during the atrocities, received her share of censure. One of the most striking of these clerical fulminations was a numerously

CLARA BARTON, PRESIDENT, AMERICAN RED CROSS,
Who is risking her life to relieve destitute Armenia.

signed and earnestly worded memorial by the Bishops of the Protestant Episcopal Church of the United States, which was presented to President Cleveland. In this document—doubtless one of the most remarkable in the history of the Christian Church in America—the reverend memorialists declare that the situation in Armenia calls for "the indignant protest of all civilized and Christian people." It then proceeds:

The entire severance of Church and State in our country should not be allowed to stifle our sympathies or hamper our action in a case like this. It should rather stimulate them. It is a case which especially appeals to us as men and Americans. As citizens of this Republic, we have learned to know and dared to maintain that no form of religious belief should expose its adherents to persecution.

It is as representatives and maintainers of this essential American principle that we appeal

THE BRITISH FLEET AT ANCHOR OFF SALONICA.

for national action in this matter of a foreign persecution whose details are too horrible to depict or enumerate.

We sincerely trust that some measure or measures consistent with the national traditions and the national dignity may be devised, and that speedily, which shall bring the whole force of the national sentiment to bear upon the solution of this subject; to cause the instant suppression of the massacres, to succor the unhappy and impoverished survivors of them, and to secure for the future ample guarantee for the safety of a Christian people in the exercise and maintenance of their faith.

We feel profoundly that our nation should cease to recognize the Turkish Government as a civilized power so long as its barbarous treatment of the Armenians continues, and that it should bring every influence to bear upon the civilized nations of Europe which may cause them to present a united front in demanding that such atrocities cease at once and forever.

Turkey having sown the wind, must reap the whirlwind, and the aftermath in the shape of claims for heavy damages that will pour in upon the Sultan's Gov-

AMERICAN MISSION AT HARPOOT, PARTIALLY DESTROYED.

ernment from many quarters, may give Abdul Hamid cause to repent some of the acts of his favorite Hamidieh troopers. Prominent among the claims to be made

THE AMERICAN COLLEGE AT MARASH, WRECKED BY KURDS AND TURKS.

in American interests are those of the American Board of Commissioners for Foreign Missions for the partial destruction of the eight buildings of the mission at Harpoot and the wrecking and looting of the handsome college at Marash. These, with similar claims for damages to the property of our citizens at many other places in Asia Minor, will be vigorously pressed. But they are probably small compared with the aggregated claims of other governments, whose citizens have suffered in person or property.

* * *

Such is the story of Eden—of that once beautiful land where, in the morning of the world, "God planted a garden," and "walked in the cool of the day," but which man's wickedness has transformed into a scene of slaughter and desolation. It may well be asked whether the Almighty has not forgotten Eden. Travelers who have passed through it recently, declare that, judging solely from its physical aspect, it would be regarded as the very last place on earth to be so favored. Treeless and barren, sterile and rocky, mountain and plain are alike uninviting; yet those bleak hills and the bare, dry valleys may have been rich in foliage and juicy grasses, while every description of flower, and shrub, and tree, luxuriant with color and laden with fragrance, may have clothed the scene with a beauty unequaled. For many centuries the human race has sought to rediscover the site of the Garden of Happiness. Scientists, explorers, historians, antiquarians and students of the ancient legend, which appears in many tongues and belongs to many lands, have searched the wide world for it. And to this high Armenian table-land the investigations of almost all have brought them at last. It meets all the requirements of Scripture and tradition. Here flowed the four rivers—the Pison, the Phrath, the Hiddekel and the Gihon, some of the ablest scholars now identifying them with the Tigris, the Euphrates, the Arras and the Djorokh rivers of to-day. "Reduced to a matter of modern geography," writes William Willard Howard, who has traveled over the entire region, "it may be said in a general way that the site of Eden is now covered by the Turkish provinces of Van, Bitlis and Erzeroum, and that the centre of the Garden would be midway between the cities of Van and Erzeroum. Included in this district are the cities of Van, Bitlis, Moush, Erzinghian and Erzeroum. The scene of the Sassoon massacre is also within the limits of the district. The caravan route from Persia to the Black Sea passes through the Garden of Eden from end to end, entering it at Baiazid and leaving it at Baiburt on the road to Trebizond." Kurds, Turks, Lazes, Circassians and Armenians dwell there, the Armenians alone being Christians, the rest their enemies and persecutors. In agriculture the land has stood still for 4000 years; in civilization it has retrograded from the patriarchal standard of early Bible times into a condition of barbarism such as no other part of the world can equal.